TECHNIQUES IN VISIBLE AND
ULTRAVIOLET SPECTROMETRY

VOLUME THREE
PRACTICAL ABSORPTION SPECTROMETRY

Practical Absorption Spectrometry

ULTRAVIOLET SPECTROMETRY GROUP

Edited by

A. KNOWLES
Department of Biochemistry,
University of Bristol

and

C. BURGESS
Glaxo Operations UK Ltd,
Barnard Castle, Co. Durham

LONDON   NEW YORK
CHAPMAN AND HALL

First published 1984 by
Chapman and Hall Ltd
11 New Fetter Lane, London EC4P 4EE

Published in the USA by
Chapman and Hall
733 Third Avenue, New York NY10017

© 1984 UV Spectrometry Group
Softcover reprint of the hardcover 1st edition 1984

British Library Cataloguing in Publication Data

Practical absorption spectrometry. –(Techniques
in visible and ultraviolet spectrometry ; v. 3)
1. Absorption spectra
I. Ultraviolet Spectrometry Group
II. Knowles, A.   III. Burgess, C.
543'.0858   QD96.A2

ISBN-13: 978-94-010-8949-4   e-ISBN-13: 978-94-009-5550-9
DOI: 10.1007/978-94-009-5550-9

Library of Congress Cataloging in Data

Knowles, A.
  Practical absorption spectrometry.

  (Techniques in visible and ultraviolet spectrometry;
  Vol. 3)
  Bibliography: p.
  Includes index.
  1. Absorption spectra. 2. Ultraviolet spectrometry.
I. Knowles, A. (Aubrey) II. Burgess, C. (Christopher)
III. Ultraviolet Spectrometry Group (Great Britain)
IV. Series.
QC459.P7   1984   535.8'4   83-15058

Contents

Preface

The inspiration for this volume lies in Edisbury's *Practical Hints for Absorption Spectrometry* which was published 17 years ago. Dr Edisbury was a founding member of the Photoelectric Spectrometry Group, served as its first Secretary and edited the Bulletin for many years. His wisdom, humour and pragmatism was evident in early meetings of the Group and in the first issues of the Bulletin, and these qualities were distilled in the writing of *Practical Hints*.

In 1977, the Committee of the Group, which by then had been re-named The UV Spectrometry Group, decided to make use of the expertise available amongst the members of the Group in writing some monographs on the practice of UV and visible spectrometry. Working parties were set up which formulated and produced the first two volumes of the series on *Standards in Absorption Spectrometry* and *Standards in Fluorescence Spectrometry*.

The success of these volumes lead the present Committee of the Group to set up a new Working Party in 1981 to plan a modern version of Edisbury's book. The idea really caught fire at the first meeting of the Working Party, when ideas sufficient to fill ten volumes were put forward. We would not pretend to emulate Edisbury's unique style, but hoped to produce a readable book for the newcomer to UV–visible absorption spectrometry, and perhaps to improve the technique of more experienced users.

The Editors had then to tackle the task of selecting from the variety of topics proposed by the Working Party and assigning the various sections of the book to its members. We hope that the result is a coherent whole that provides the kind of information that is not found in standard texts or in instrument manufacturers' literature. We have attempted to survey the types of instrument that are currently in use, and hope that this provides assistance in the selection of an instrument from the bewildering range that is now available.

We decided that it would be invidious to provide a comprehensive *Which* report and have as far as possible avoided mentioning instruments by name: this is in accord with the Group's policy of impartiality. Where instruments or components are named, this does not imply that the UV Spectrometry Group in any way endorses that particular instrument or component.

The euphoria of completing the manuscript has been marred by the untimely death of Dr Brian Chadburn, who was a member of the Working Party and Honorary Treasurer of the UV Group. We gratefully acknowledge his enthusiastic support of the project, and will miss his participation in the future activities of the Group.

We have received support and assistance from many people outside the Working Party. In particular, we would like to thank Messrs Bausch and Lomb, Beckman, Hewlett Packard, Perkin—Elmer, Pye Unicam, Shimadzu, Starna and Thermal Syndicate for permission to reproduce diagrams. Our thanks are due to Anne Grundy, Benita Hall, Wendy McElroy, Jillian Wearmouth and Jennifer Wood for help with the typing, and to Bob Bourne and Linda Stanley for running spectra. We also thank Dr T.L. Threlfall for assistance with Chapter 1, and Susan Donaghy and Mary Ann Ommanney of Chapman and Hall for their support and encouragement. All proceeds from sales of this book will go to Group funds for the furtherance of UV spectrometry.

February 1983   AUBREY KNOWLES
                                       CHRIS BURGESS

Membership of the working party

Mr J. G. Baber, Glaxo Operations (UK) Ltd, Barnard Castle DL12 8DT, Co. Durham

Dr W. S. Brickell, Glaxo Group Research Ltd, Greenford UB6 0HE, Middlesex

Dr C. Burgess, Glaxo Operations (UK) Ltd, Barnard Castle DL12 8DT, Co. Durham

Dr B. P. Chadburn,† Perkin—Elmer Ltd, Post Office Lane, Beaconsfield HP9 1QA, Buckinghamshire

Dr A. F. Fell, Department of Pharmacy, Heriott-Watt University, 79 Grassmarket, Edinburgh EH12 2HJ

Dr M. A. Ford, Perkin—Elmer Ltd, Post Office Lane, Beaconsfield HP9 1QA, Buckinghamshire

Mr D. Irish, Pye Unicam Ltd, York Street, Cambridge CB1 2PX

Dr A. Knowles, Department of Biochemistry, University of Bristol, Bristol BS8 1TD

Dr W. F. Maddams, BP Research Centre, Chertsey Road, Sunbury-on-Thames TW16 7LN, Middlesex

Dr M. A. Russell, BDH Chemicals Ltd, West Quay Road, Poole BH12 4NN, Dorset

Dr R. L. Tranter, Glaxo Pharmaceuticals Ltd, Barnard Castle DL12 8DT, Co. Durham

Mr P. W. Treherne, Bausch and Lomb UK Ltd, Wingate House, Wingate Road, Luton LU4 8PU, Bedfordshire

Dr J. F. Tyson, Department of Chemistry, University of Technology, Loughborough LE11 3TU, Leicestershire

† Deceased

Glossary

Terms and abbreviations used in absorption spectrometry

Full definitions and examples of most of the terms will be found by reference to the index. In general, SI units are employed, but in several instances older units are still used by most spectroscopists. Some abbreviations used in the book are not listed here, but will be found in the Index.

A. See Absorbance.

$A_{1cm}^{1\%}$. See $E_{1cm}^{1\%}$

Å. See Ångstrom unit.

Absorbance. Quantity expressing the absorption of radiation by a solution at a specified wavelength. It is given by:

$$A = \log 1/T = -\log T$$

and is linearly related to the pathlength and concentration of the solution. It is dimensionless but is expressed in absorbance units (A) so that a solution of $T = 0.1$ has an absorbance of 1 A.

Absorption. The process by which radiation is attentuated on passing through a substance. The term implies that the radiant energy is converted into some other form, e.g. heat, fluorescence etc., as distinct from losses by scattering or refraction.

Absorption band. See Band.

Absorption spectrum. A plot of the absorption of radiation by a sample against the wavelength of the radiation.

Absorptivity. The absorbance of a solution of a compound in unit concentration measured in unit pathlength at a specified wavelength. See Molar absorptivity.

Ångstrom unit (Å). A unit of wavelength, now rarely used in absorption spectrometry. 1 Å = 0.1 nm = 10^{-10} m.

b. See Pathlength.

Band. A general term describing a maximum in a plot of some quantity against wavelength. An absorption band is a broad maximum in an absorption spectrum and may comprise a number of minor peaks. An emission band might be the profile of an intensity versus wavelength plot of the spread of wavelengths leaving a monochromator.

Band pass. See Passband

Beer's Law (Beer–Lambert–Bouguer Law). This relates the absorbance of a solution to the pathlength of the cell and the concentration of the solute. The absorbance of a solution at a specified wavelength is given by:

$$A = \epsilon bc$$

where ϵ is the molar absorptivity at that wavelength, b is the pathlength in cm and c is the molar concentration.

Blazed grating. A diffraction grating made with a particularly high reflectivity in a particular spectral region.

c. Velocity of light.

cm^{-1}. See Wavenumber.

cps. Cycles per second; see Hertz.

Cell. A container to hold solutions for the measurement of their absorption spectra. This should have two parallel optical windows fixed at a specified distance apart.

Coated optics. The coating of optical components to improve their optical performance and protect them against atmospheric attack.

Cuvette. See Cell.

Dark current. The background signal from a detector due primarily to the thermal emission of electrons.

Derivative spectra. A plot of the first, second or higher derivative of an absorbance spectrum with respect to wavelength in order to correct for background absorption and to accentuate minor features in the spectrum.

Detector. The device in a spectrometer which measures the intensity of light transmitted by a sample.

Deviation. The displacement of the spectrometer measuring beam due to optical defects in the cell, or in its alignment.

Difference spectrum. Small changes in the absorbance spectrum of a sample can be more readily detected if the spectrum is measured using the original sample as a reference solution.

Diffraction grating. A device used to resolve radiation into its component wavelengths by an interference process. See Grooves.

Dispersion. The power of a prism or grating to separate radiation of different wavelengths.

Double-beam spectrometer. An instrument in which the measuring beam is split into two equivalent paths, one passing through the sample and one through a reference cell.

Dual-wavelength spectrometry. Measurement of the absorbance of a sample at two wavelengths simultaneously in order to compensate for background absorption, etc.

$E_{1\,cm}^{1\%}$. An expression of the absorptivity of a solute of unknown molecular weight. It is the absorbance of a 1% w/v solution of a compound measured in 10 mm pathlength. It is related to molar absorptivity by:

$$E_{1\,cm}^{1\%} = 100 \times \epsilon/M.$$

ESW. See Effective spectral slitwidth.

Effective spectral slitwidth (ESW). The band of wavelengths emerging from a monochromator, expressed as the bandwidth at half-peak height.

Electromagnetic radiation. Radiation which can be regarded as wave motions of characteristic wavelength, including γ-radiation, X-rays, UV, light, IR and radio waves.

Emission. The process by which electromagnetic radiation is radiated from a substance.

Emission line. Emission of radiation in a very narrow wavelength band characteristic of the output of gas discharge lamps.

Excited state. A molecule that has absorbed UV or VIS radiation is said to have been raised from its ground state to an excited state.

Extinction. See Absorbance.

Extinction coefficient. See Molar absorptivity.

Far stray-light. The component of the monochromator stray-light lying at wavelengths well away from the passband. This may be of low intensity but can be a major factor in the problem of instrumental stray-light.

Far-ultraviolet. Radiation at the shortwave end of the UV region. There are no agreed limits, but the far-UV is generally taken to lie between 190 and 250 nm.

First order. See Grating order; Derivative spectra.

Flow cells. A cell designed for the measurement of a flowing stream of solution.

Fluorescence. A process in which radiation absorbed by a molecule is instantaneously re-emitted as UV or VIS radiation. This is usually of longer wavelength than the exciting radiation.

Fluorimeter. An instrument specifically intended for the measurement of fluorescence.

Frequency (ν). A specification of electromagnetic radiation by the number of waves per unit time. It is related to wavelength by:

$$\nu = \lambda/c \text{ Hz}$$

where λ is the wavelength in m and c is the velocity of light in m s^{-1}.

Gaussian. Most absorption bands when plotted on a frequency scale have a shape equivalent to a Gaussian or normal distribution, and can be fitted by:

$$A = A_m \exp -2.773((\nu - \nu_m)/w)^2$$

where A is the absorbance at frequency ν, A_m is the absorbance at the maximum, frequency ν_m, and w is the width of the band at $0.5 A_m$. Also see Lorentzian.

Grating. See Diffraction grating.

Grating order. A diffraction grating will reflect a given wavelength of light at more than one angle. Most of the incident energy undergoes first-order reflection, the second-order reflection is weaker and so on.

Grooves. A diffraction grating works by the interference of light rays reflected from the faces of grooves ruled in its surface. The spacing of these grooves determines the optimum wavelength range for the grating.

Hz. See Hertz.

Hertz. Unit of frequency equivalent to one wave or cycle per second.

Holographic grating. A diffraction grating produced photographically by the interference of beams of light.

IR. See Infrared.

ISL. See Instrumental stray-light.

Incident beam. The radiation directed into the entrance window of the cell in the spectrometer.

Infrared. The region of electromagnetic radiation of wavelength greater than the red end of the VIS region, i.e. from 800 nm to 1000 μm.

Instrumental stray-light (ISL). The overall effect of stray-light in a spectrometer manifested as detector output. Thus it is the signal generated in the detector by all wavelengths outside the monochromator passband that reach it.

Isosbestic point. If a sample contains two compounds in equilibrium, then at any wavelength at which their molar absorptivities are the same, the absorbance of the sample will be independent of the relative amounts of the two species. When the spectra of a series of mixtures of differing compositions are plotted, these will be seen to cross at these wavelengths, forming isosbestic points.

Light. See Visible light.

Lorentzian. Some absorption bands when plotted on a frequency scale have a shape equivalent to a Lorentzian or Cauchy distribution and can be fitted by:

$$A = A_m / \{1 + [2(\nu - \nu_m)/w]^2\}$$

where A is the absorbance at frequency ν, A_m is the absorbance at the maximum frequency ν_m, and w is the width of the band at $0.5 A_m$. Also see Gaussian.

Luminescence. A general term for the emission of UV or VIS radiation from a molecule. In most cases, it can be regarded as the sum of the fluorescence and the phosphorescence emissions.

MSL. See Monochromator stray-light.

mμ. Millimicron; see Nanometre.

Manual spectrometer. An instrument without a wavelength drive mechanism.

Mask. An opaque screen with an aperture designed to limit the cross-section of a light beam.

Microcell. A cell with a narrow chamber requiring the minimum volume of sample for a given pathlength.

Micron. Micrometre (μm), 10^{-6} m.

Millimicron. Nanometre (nm), 10^{-9} m.

Molar absorptivity (ϵ). The absorbance at a specified wavelength of a solution of a compound of unit molar concentration measured in a 10 mm pathlength. It has dimensions of M^{-1} cm^{-1}.

Monochromator stray-light (MSL). Radiation emerging from a monochromator of wavelengths outside the passband to which the monochromator is set.

NBW. See Natural bandwidth.

NIR. See Near-infrared.

Natural bandwidth (NBW). The width of an absorption band of a particular substance, measured at half-peak height.

Near-infrared. The part of the infrared region closest to the VIS region, generally taken to be between 800 nm and 2 μm.

Near stray-light. The component of the monochromator stray-light lying within a few nanometres of the monochromator passband.

Near-ultraviolet. The part of the UV region closest to the VIS region. There are no agreed limits, but the range is generally taken to be 250 to 400 nm.

OD. See Optical density.

Optical density. This is a general term for the absorbance of any material: absorbance should only be applied to solutions.

Passband. The band of wavelengths emerging from a monochromator, generally centred on the indicated wavelength.

Pathlength. The thickness of a sample solution traversed by the beam as it passes through a cell.

Phosphorescence. A process in which radiation absorbed by a molecule is re-emitted as UV or VIS radiation, generally of longer wavelength than the exciting radiation. Unlike fluorescence, there is a delay between absorption and emission.

Photon. A quantum of UV or VIS radiation.

Phototube. A vacuum tube (valve) detector.

Photovoltaic detector. A detector which generates a voltage in response to light.

Quantum. A fundamental unit of radiation. The energy of a quantum is related to the frequency of the radiation by:

$$E = h\nu$$

where h is Planck's constant.

Radiation. See Electromagnetic radiation.

Reference cell. A cell identical to that used for the sample solution, but containing only solvent. The difference between the absorbances of sample and reference cells is thus a true measure of the absorbance of the solute.

Resolving power. The ability of an instrument to distinguish between two closely-spaced maxima.

SSW. See Spectral slitwidth.

Sampling cell. A cell designed to measure a number of discrete samples in succession, each sample displacing its predecessor.

Scan. To drive a spectrometer through its wavelength range in order to measure an absorption spectrum.

Scatter. See Scattered light.

Scattered light. Radiation which is reflected or refracted out of the measuring beam during its passage through the spectrometer. This may occur due to optical defects, dust or contamination of the optical surfaces in the instrument, or due to particles or inhomogeneities in the sample solution.

Second order. See Grating order; Derivative spectra.

Semi-micro cell. A cell with chamber of reduced width and thus requiring less sample solution to fill it.

Single-beam spectrometer. An instrument with a single optical path. This means that for each wavelength setting, the sample cell must be moved out of the beam so that the transmission scale can be set to unity.

Sipper system. A manual or electric pump system used to draw sample solutions into a sampling cell.

Slew rate. A term used to describe both the rate of wavelength scan of a spectrometer and the rate of movement of the recorder pen.

Slit. An opaque screen with a fixed or adjustable aperture designed to limit the width of a light beam.

Slitwidth. The width of the aperture of a slit. Monochromator slits determine both the amount of light and the spread of wavelengths transmitted by the monochromator.

Spectral line. Generally synonymous with emission line, but can be used for very narrow absorption maxima.

Spectral slitwidth (SSW). The range of wavelengths emerging from a monochromator when it is set to a particular wavelength and slit opening. Effective spectral slitwidth is the preferred means of expressing this.

Spectrometer. An instrument for measuring the transmittance or absorbance of a solution at different wavelengths.

Spectrophotometer. See Spectrometer.

Spectrum. See Absorption spectrum.

Stray-light. Radiation present in a spectrometer beam of wavelengths outside the monochromator passband. This may be due to optical defects, dust etc.

T. See Transmittance.

Transmission. The process by which radiation passes through a material. It therefore represents radiation that is not absorbed, scattered or otherwised dispersed by the material.

Transmittance. The proportion of light that is transmitted by a sample:

$$T = I/I_0 = 10^{-A} = \text{antilog}(-A)$$

where I_0 and I are the intensities of the light falling on the sample and that emerging from it. Thus T lies in the range from 0, for an opaque material, to 1 for a transparent one. T has no units, but is sometimes expressed as a percentage, i.e.

$$T\% = 100 \times I/I_0$$

UV. See Ultraviolet.

UV–VIS spectrometer. An instrument designed to operate through the UV and VIS regions, i.e. from 180 to 800 nm.

Ultraviolet (UV). The region of the electromagnetic spectrum lying at shorter wavelengths than the blue end of the visible region, and generally taken to be between 100 and 400 nm.

VIS. See Visible light.

Vacuum ultraviolet. The shortwave part of the UV region where oxygen and other gases absorb strongly, lying between 100 and 180 nm. Spectrometers for this region must be housed in evacuated chambers.

Visible light. The region of electromagnetic radiation that can be seen by the human eye, and generally taken to extend from 400 nm (violet) to 800 nm (red).

Wavelength (λ). A measure of electromagnetic radiation, being the length of the waves associated with the radiation. In the UV–VIS region, these waves are very small being of the order of 10^{-7}–10^{-6} m in length; about 150 waves of green light would span the thickness of this page.

Wavenumber ($\bar{\nu}\,\text{cm}^{-1}$). An alternative measure of frequency, expressed as the number of waves cm^{-1}. This can be derived from wavelength without knowledge of the velocity of light:

$$\bar{\nu} = 10^7/\lambda \ \text{cm}^{-1}$$

where λ is the wavelength in nanometres.

Window. The optical faces of a cell.

Working area. That part of the window of a cell that is up to optical specification and through which the measuring beam can pass without risk of interference with the walls or floor of the cell, or with the meniscus of the sample.

Zero order. In discussing derivative spectra, zero order is used to describe the fundamental absorption spectrum.

α. Sometimes used for absorptivity or molar absorptivity.

ϵ. Molar absorptivity.

ϵ_{max}. The maximum absorptivity of an absorbance band.

λ. Wavelength.

λ_{max}. The wavelength of maximum absorbance of an absorption band.

μm. Micrometre.

ν. Frequency.

$\bar{\nu}$. Wavenumber.

1 Absorption spectrometry

1.1 Absorption spectrometry in the ultraviolet and visible regions

It is now 35 years since the first spectrometers working in the ultra-violet and visible (UV−VIS) regions of the spectrum came into general use, and over this period they have become the most important analytical instrument in many chemical, biological and clinical labor-atories. Because the technique has become so commonplace, it is assumed that every scientist knows how to 'run' an absorption spectrum. However, a proper training in the technique is essential for there are many pitfalls to be avoided if reliable results are to be produced. Although this book is intended to be an introduction to those new to the technique, it may also help to improve the results of more experienced users.

The potential of absorption spectrometry is best described by out-lining some of its merits and limitations.

1.1.1 *Characterization of compounds*

Most organic compounds and many inorganic ions and complexes absorb radiation in the UV−VIS region. A plot of this absorption by a compound against wavelength is called its *absorption spectrum*; this has a shape that is characteristic of a particular compound or class of compounds. The UV−VIS spectrum does not usually give enough information to identify an unknown compound, but when combined with other analytical techniques or with chromatographic separation, an unequivocal identification can be achieved. Absorption spectrometry is a non-destructive technique and is extremely sensitive, and is therefore ideal for the characterization of small amounts of precious compounds. As an extreme example, it is used for the measurement of the pigments in the single retinal receptor cells. A custom-built-microspectrometer is used which is capable of recording satisfactory spectra from only 4×10^{-16} mol pigment, which is only about 10^8 molecules.

1.1.2   *Quantitative assay*

The most important application of the technique is as a means of measuring concentration and modern instruments are designed to facilitate rapid and accurate measurements. The precision that can be achieved depends upon a number of factors that will be discussed later in this volume, but single measurements of precision better than $\pm 0.5\%$ should be possible. One particularly important merit of the technique is that trace components can be measured in the presence of high concentrations of other components if there is sufficient difference in their absorption spectra. Similarly, mixtures of compounds with differing spectra can be analysed, and methods for doing this will be discussed in Chapter 10.

The technique is best suited to dilute solutions, though gases and solids can be measured by special methods. Thus solubility in a suitable solvent is a prerequisite for the accurate measurement of a particular sample. The optimum concentration range for the measurement of a compound is limited and does not generally exceed a 100-fold range, and so it may be necessary to adjust the concentration of the test solution to bring it into the optimum range. Beyond manipulations of the concentration of solutions, measuring the UV–VIS absorption will not change the sample in any way and even UV-sensitive compounds will suffer negligible damage, for the amount of radiation absorbed by the sample during the measurement is very small.

1.1.3   *Rapid assays*

The speed of photodetectors and of modern electronic components means that the measurement of solutes in flowing or rapidly changing systems can be carried out with millisecond time resolution. A popular application is the monitoring of the eluent from HPLC columns: the concentration of a particular component can be followed by measuring the absorbance at a suitable wavelength, or compounds eluted from the column can be identified by making rapid spectral scans.

Absorption measurements are also the most popular means of following the kinetics of reaction systems since they do not interfere with the progress of the reaction in any way. The 'stopped flow' technique is an important method for the study of biological reactions, and light absorption is the most popular monitoring technique since it has both speed and selectivity for a particular component of a reaction mixture. Probably the most advanced kinetic monitoring systems are found in flash photolysis systems where reactions with lifetimes of the order of 10^{-8} s can be followed. By using very fast detection systems, such as streak cameras, absorption measurements in the picosecond time domain are possible. Another important application of rapid absorption measurement is in the clinical field, where colorimetric assays have been worked out for many biologically important compounds. Continuous flow systems employing these reactions can measure many hundreds of samples in an hour.

All of these techniques, however advanced, are based on the same fundamental principles of spectrometry and a sound grasp of these particular principles is essential, whatever kind of measurement you intend to make. The remainder of this chapter will introduce the fundamentals of absorption spectrometry, and the rest of the volume will be concerned with the practicalities of their application.

1.2   The ultraviolet and visible spectrum

Fig. 1.1 shows the electromagnetic spectrum and the narrow region that is of interest to us. Despite its narrowness, this band of radiation is vital to life on earth because its interactions with molecules is the primary step in both photosynthesis and vision. Measurement of such molecular interactions forms the basis of UV–VIS spectroscopy and can provide a wealth of information about the molecules. The diagram shows that radiation can be measured in terms of either wavelength or frequency. In the UV–VIS region, wavelength is generally used, even though frequency would be more appropriate. The convenient unit of wavelength in this region is the *nanometre* (nm, 10^{-9} m, formerly called the millimicron, mμ). When frequency is used, it is generally expressed in terms of *wavenumbers* – the number of waves per cm – rather than the number of waves per unit time, thus avoiding any assumptions about the speed of light in a particular medium. These quantities are related by:

$$\lambda = \frac{10^9 c}{\nu} = \frac{10^7}{\bar{\nu}}$$

where λ is the wavelength in nm, ν is the frequency in Hz, $\bar{\nu}$ is the frequency in wavenumbers (cm^{-1}) and c m s^{-1} is the velocity of light (2.998×10^8 m s^{-1} in air).

The limits of the *visible spectrum* are ill-defined but most instru-

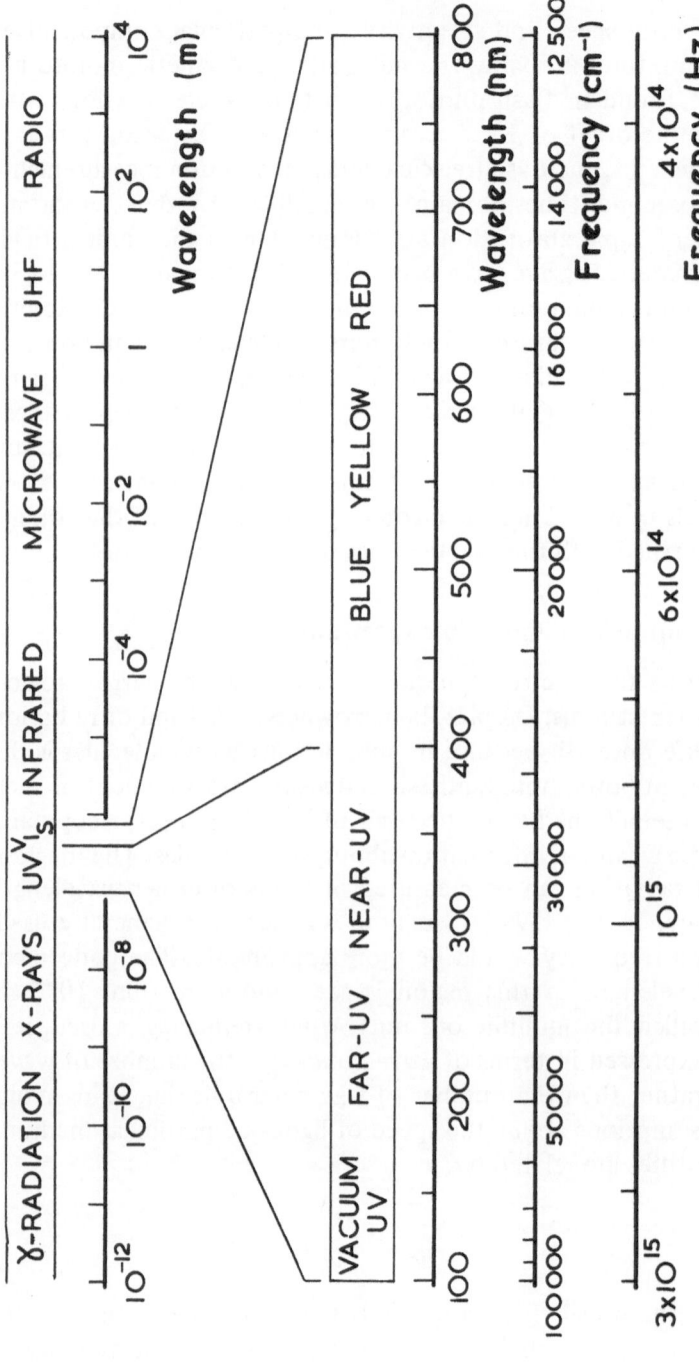

Fig. 1.1 *The range of electromagnetic radiation. The upper part is plotted on a logarithmic scale and illustrates the small extent of the UV–VIS region. The lower part is an enlargement of this region on a linear wavelength scale and shows the relationship of this scale to the frequency scales.*

ment makers take it to lie between 400 and 800 nm; only radiation lying between these limits can properly be described as 'light'. The *ultraviolet region* extends from 400 nm down to 100 nm. Since most gases have appreciable absorption below 185 nm, measurements in the range below this wavelength can only be made using instruments in evacuated enclosures; this *vacuum-UV* region is therefore outside the range of most instruments and will not be dealt with in this book. The remainder of the UV region is loosely divided into the *near-UV* and *far-UV* regions, though there is no consensus on where the dividing line should be, and different authorities place it at various points between 200 and 300 nm. These terms will only be used in a relative sense in this text.

1.3 The absorption of radiation

When UV or VIS radiation encounters an atom or molecule, an interaction between the radiation and the electrons of the latter may take place. This absorption process is very specific and results in an attenuation of the radiation and an increase in the energy of the electrons of the atom or molecule. This may be regarded as the promotion of one of the outer or bonding electrons from a 'ground-state' energy level into one of higher energy (Fig. 1.2). These levels are separated by a discrete energy increment, E, which is determined by the nature of the atom or molecule, and only parcels of radiation of energy E can be absorbed. This parcel of radiation is termed a *quantum* and

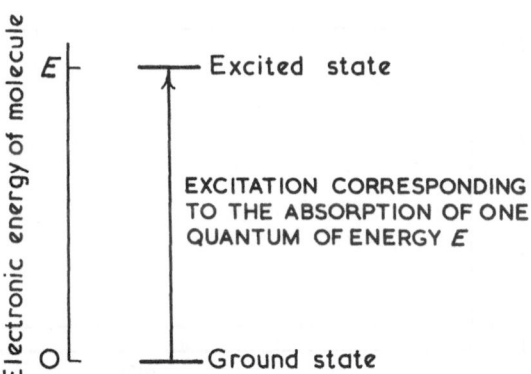

Fig. 1.2 *A representation of the absorption of radiation by a molecule. This results in the excitation of one electron from the ground-state to an excited state of energy E.*

its energy is related to the frequency and wavelength of the radiation by:

$$E = h\nu = \frac{hc}{\lambda} \times 10^9$$

where h is Planck's constant (6.63×10^{-34} J s), c is the velocity of light (2.998×10^8 m s^{-1}) and λ is in nm.

Suppose that Fig. 1.2 represents the energy level diagram of a molecule and that $E = 7.95 \times 10^{-19}$ J which is equivalent to $\lambda = 250$ nm. If the molecule is exposed to a complete spectrum of UV and VIS radiation, only that of wavelength exactly 250 nm will be absorbed. A plot of absorption versus wavelength – the *absorption spectrum* – would be a single sharp line as shown in Fig. 1.3. In reality even the simplest molecules have large numbers of energy levels and their absorption spectra are far more complex than this. In addition, each electronic energy level has a group of closely spaced vibrational levels associated with it due to small increments of the energy of the molecule caused by the relative motions of its constituent atoms. These vibrational levels overlap to such an extent that most spectrometers are incapable of resolving them and the measured spectrum appears as a broad bell-shaped peak.

Benzene is a molecule comprised of 12 atoms and its absorption spectrum is relatively complex. Fig. 1.4a shows a part of the UV spectrum and has a group of minor peaks that form an *absorption band* representing a major type of electronic excitation. The minor peaks represent small differences in the energies of electronic excitations within the group, while the rounded shape of each of these

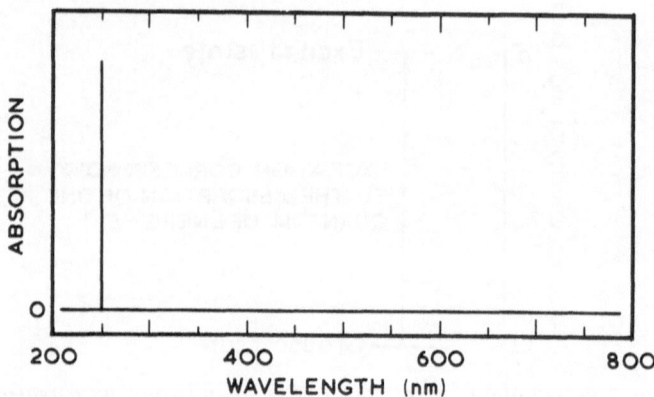

Fig. 1.3 *The hypothetical absorption spectrum for a molecule having the single absorption process shown in Fig. 1.2.*

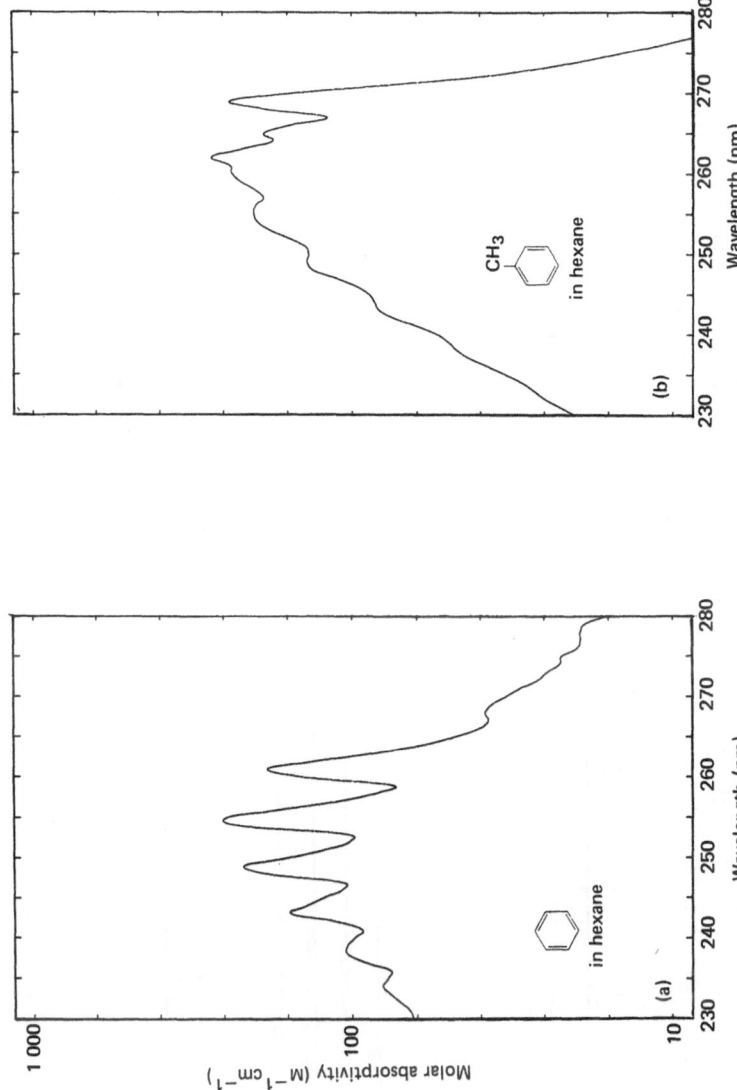

Fig. 1.4 *UV absorption spectra of (a) benzene and (b) toluene dissolved in hexane.*

minor bands is due to the fusion of the myriad vibrational levels associated with each electronic transition. The composition of these minor bands becomes more apparent if the benzene spectrum is measured with the sample in gaseous form rather than in solution. Fig. 1.5 shows the 240−265 nm region of the vapour-phase spectrum run on a conventional spectrometer under conditions of maximum resolution. This shows how each of the four maxima seen in this region in the solution spectrum can be resolved into a major peak followed by a series of lesser ones of decreasing energy. Even this vapour-phase spectrum cannot be regarded as an 'absolute' spectrum of benzene, for an ultra-high-resolution spectrometer would be capable of further separating these peaks into even narrower ones.

Although appearing very complex, the benzene spectrum is relatively simple because of the great symmetry of the molecule. The majority of organic compounds have so many overlapping bands that these merge into one or two broad maxima when their spectra are measured in solution. The spectrum of aniline (Fig. 1.6a) is thus more typical of organic compounds in general and also serves to illustrate how simple substitution of the benzene molecule can transform its spectrum.

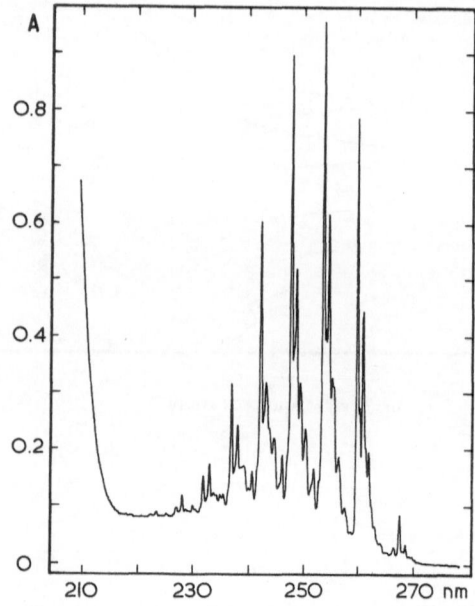

Fig. 1.5 *An absorption spectrum of benzene vapour. Air was saturated with vapour at 1 atm and measured in 5 mm pathlength. ESW about 0.03 nm.*

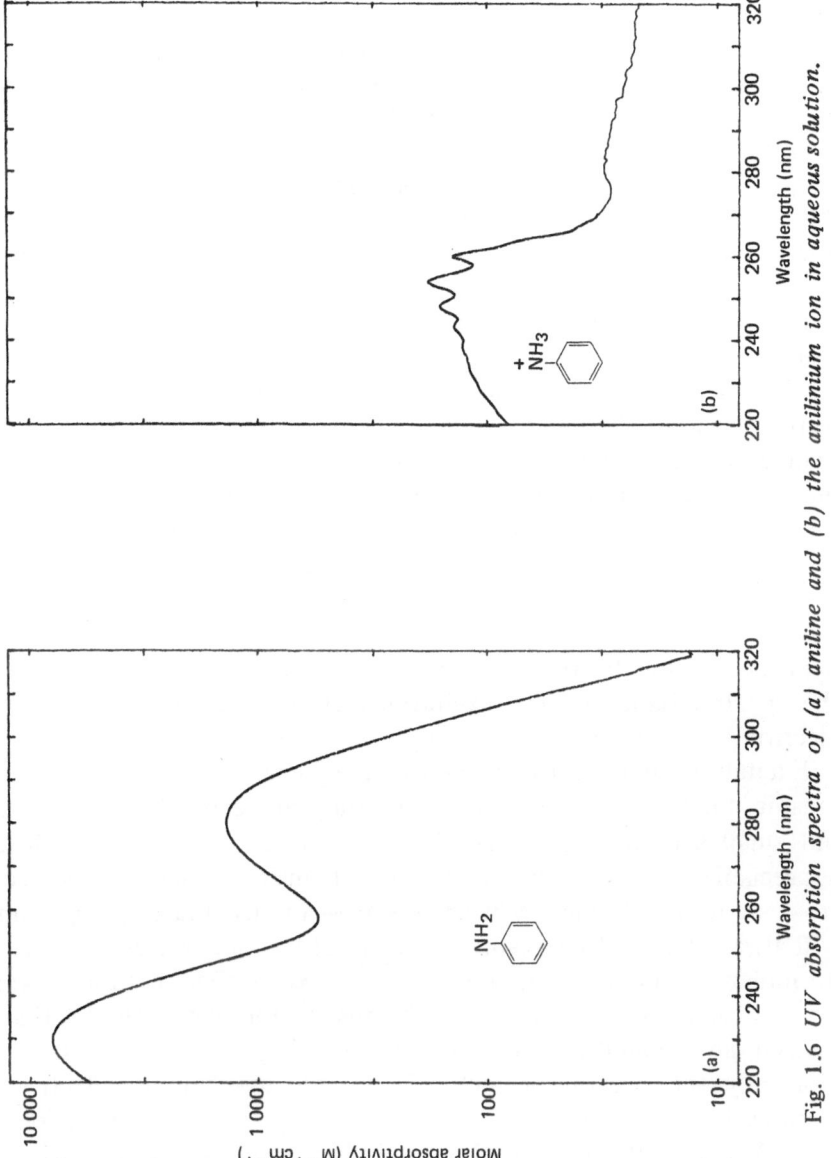

Fig. 1.6 UV absorption spectra of (a) aniline and (b) the anilinium ion in aqueous solution.

1.4   Molecular structure and absorption spectra

This is a vast subject that can only be briefly outlined here. Despite the great amount of effort that has been put into the prediction of absorption spectra by the calculation of molecular energy levels, most discussions of structure–spectra correlations are based on empirical rules, and these are always hedged around by provisos and qualifications. Most molecules absorb somewhere in the UV–VIS region and, in general, the more complex the molecule, the longer the wavelength of its first absorption band – that is, the band of lowest energy and longest wavelength. Thus the simplest molecules, for example, O_2, absorb appreciably only below 190 nm, while the first band of a complex molecule like methylene blue lies at the red end of the visible region.

The relationship of the absorption spectra of organic compounds to their structure has been extensively studied and is, to some extent, understood. The first step in predicting the absorption spectrum of a molecule is to consider its bonding electrons. The outer electrons of organic compounds are of three main types: σ-electrons which are involved in covalent bonds, π-electrons which are involved in double and triple bonds, and n-electrons which are the non-bonding electrons associated with hetero atoms such as nitrogen and oxygen. Current explanations of absorption spectra are based on the assignment of the spectral bands to the excitations of these different classes of electron.

Excitation of σ-electrons requires the highest energies. Since all organic compounds have σ-electrons, they all absorb UV radiation, but small saturated molecules like the smaller alkanes only show appreciable absorption below 200 nm. Transitions involving π-electrons are found at longer wavelengths, thus an unsaturated compound will always have a first absorption band at a longer wavelength than a similar saturated compound. This is exemplified by Fig. 1.4a; hexane has no π-electrons and so absorbs at shorter wavelengths than benzene and could thus be used as non-absorbing solvent for the latter. Not only the size of the molecule but the number of π-electrons forming the conjugated system has a profound effect upon the location (λ_{max}) of the first band. This is illustrated in Fig. 1.7, where increasing the number of conjugated rings is seen to displace the first band from the UV into the visible region, naphthacene appearing coloured.

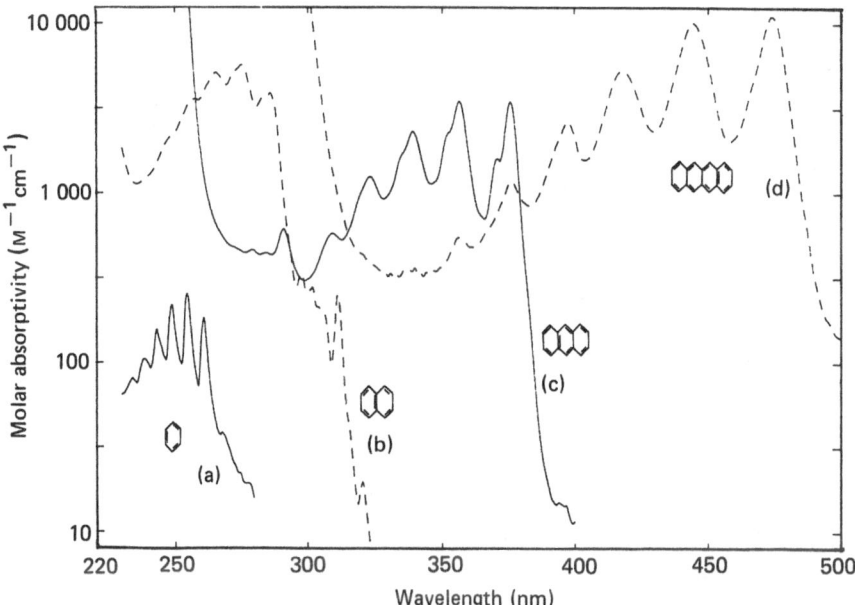

Fig. 1.7 *Absorption spectra of (a) benzene, (b) naphthalene, and (c) anthracene in hexane, and (d) naphthacene in toluene.*

1.4.1 *Substituent effects*

The absorption spectra of compounds formed from a hydrocarbon by the simple substitution of hydrogen atoms by other groups can to some extent be predicted, since a given substituent has a known effect upon the electrons of the molecule and consequently its absorption spectrum. For example, alkyl groups have only minor effects upon the electronic orbitals of a molecule and hence little effect upon the absorption bands. Fig. 1.4b shows the absorption spectrum of toluene: comparison with Fig. 1.4a shows that the absorption bands of benzene have been smoothed by the introduction of further minor energy levels, and the centres of the bands have been shifted to longer wavelength. This displacement is described as a *bathochromic shift*. Substituents containing hetero atoms generally have greater effects than alkyl substituents, as illustrated by the spectrum of aniline (Fig. 1.6a). In addition, if these atoms can take up or lose an electron, further major changes in the absorption spectrum are seen. Thus if an aniline solution is acidified, a proton is taken up to form the anilinium ion (Fig. 1.6b). The effect of tying up the nitrogen lone pair electrons in this way is to shift the bands to shorter wavelengths (an *hypso-chromic shift*) and the spectrum becomes very similar to that of toluene.

Substituents containing C=C, C=N or C=O double bonds in positions where they can conjugate with double bonds in the core molecule will cause much larger hypsochromic shifts than alkyl substituents, and may also introduce extra characteristic bands of their own. Thus many carbonyl compounds have a band at about 280 nm due to the excitation of an oxygen lone pair electron into an unoccupied π-orbital: It is therefore designated an $n \rightarrow \pi^*$ transition. This band is the main feature in the spectrum of butanone shown in Fig. 1.8. The dominance of each characteristic band in the spectra of small molecules can be of help in classifying an unknown compound, but it is not generally possible to distinguish between compounds in a given class. For example, the spectra of the naturally-occurring pyrimidine bases are very similar and cannot be used to make positive identification. However, in this case, individual compounds can be characterized by means of their pK, and this can be conveniently determined by UV absorption measurements. Ionization of the compounds results in a major spectral change (Fig. 1.9) and so measurements of the spectra over a range of pH values will enable the pK to be accurately evaluated.

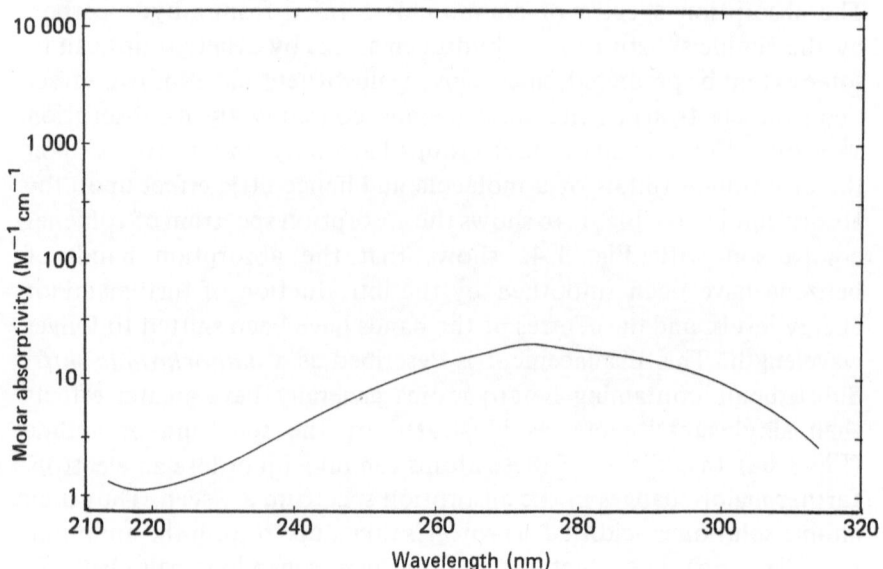

Fig. 1.8 *UV absorption spectrum of butan-2-one in heptane.*

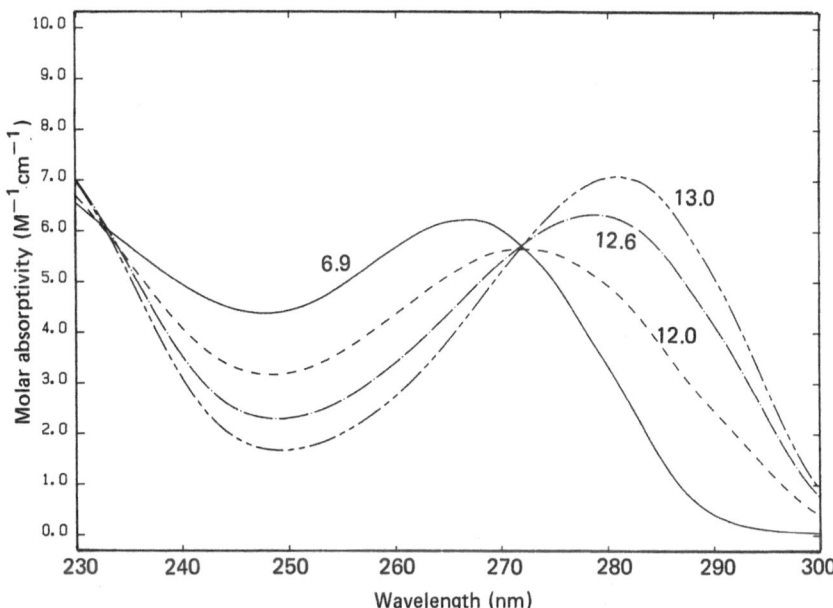

Fig. 1.9 *The absorption spectrum of an aqueous solution of cytosine at the pH values shown. The spectra at pH = 6.9 and 13.0 are taken to represent the proton- ated and unprotonated molecules, respectively, and the spectra at intermediate pH represent mixtures of the two species. Both species have the same molar absorptivity at about 234 and 272 nm; consequently the spectra all cross at these wavelengths forming isosbestic points. From a series of such curves, the pK_a can be shown to be 12.2.*

1.4.2   Solvent effects

A change in the pH of the medium generally has a profound effect upon the spectrum of a molecule containing hetero atoms. Beyond this effect, a change in the polarity of the solvent can cause smaller shifts in the spectra of most compounds. An increase in solvent polarity will cause a bathochromic shift of $\pi \rightarrow \pi^*$ absorption bands, though not all of the band systems in a particular molecule may be affected to the same extent and so the shape of the spectrum may well change. $n \rightarrow \pi^*$ transitions are also sensitive to solvent polarity, but undergo hypsochromic shifts when the polarity is increased. Table 1.1 illustrates this effect on the 280 nm absorption band of a typical carbonyl compound; other solvent properties beyond polarity – measured as dielectric constant – are involved in the observed shifts, for example, the ability of the solvent molecules to form hydrogen bonds, but the relationship is evident. Since the effect of

Table 1.1: *The effect of solvent polarity upon the* $n \rightarrow \pi^*$ *absorption band of acetone. Data from [1].*

Solvent	Dielectric constant	λ_{max} (nm)	Hypsochromic shift $\Delta \bar{\nu}$ (cm^{-1})
Hexane	1.9	280	0
Carbon tetrachloride	2.2	280	0
Dioxane	2.2	277	300
Chloroform	4.8	276	400
Dimethylformamide	38.4	275	500
Dimethyl sulphoxide	44.6	274	600
Ethanol	24.3	271	1000
Methanol	32.6	270	1200
Acetic acid	6.2	267	1600
Water	78.5	265	2000

changing solvent differs for $\pi \rightarrow \pi^*$ and $n \rightarrow \pi^*$ bands, compounds having both types of bands will show major changes in their spectral shapes. The possibility of solvent effects must be remembered when comparing spectra measured in the two common spectroscopic solvents, hexane and ethanol, which differ greatly in polarity. As Table 1.1 shows, the absorption maximum of acetone is shifted by 9 nm on going from one to the other.

1.5   Quantitative absorption spectrometry

Beyond having characteristic peak positions, the absorption spectrum of a given compound also has characteristic peak heights which can serve as an additional aid to identification and – more important – will form the basis of a quantitative assay. When radiation travels through the solution of an absorbing compound it is reduced in intensity by each molecule that it encounters according to an exponential law. The amount of radiation absorbed by a solution is thus an exponential function of the concentration of the solution and the distance that the radiation passes through it (Fig. 1.10). In practice it is the amount of radiation transmitted by the solution that is measured. This is expressed as the *transmittance*, which is the ratio of the amount of radiation transmitted to that incident upon the solution:

$$T = \frac{I}{I_0} \propto e^{-bc}$$

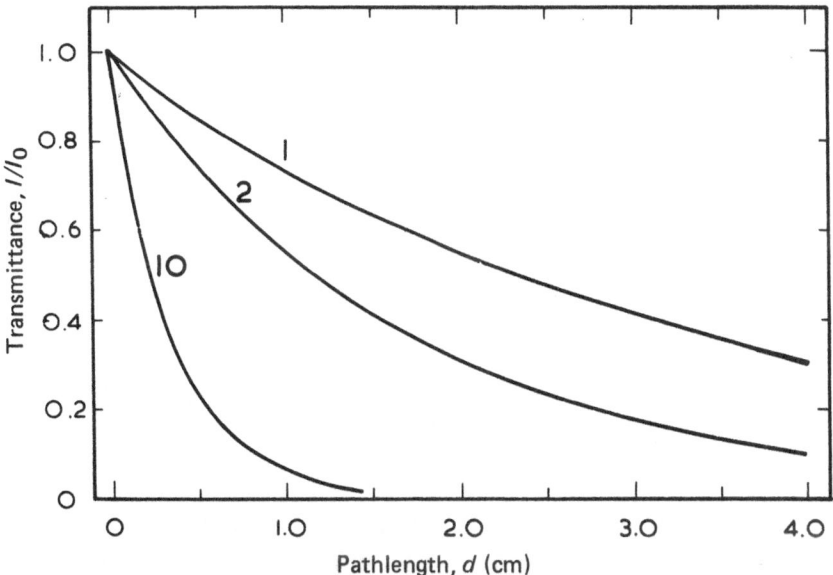

Fig. 1.10 *The absorption of radiation on passing through solutions of different thicknesses and concentrations. The curves show the attenuation of the measuring beam when passing through solutions of an absorbing solute at relative concentrations of 1, 2 and 10.*

where T is the transmittance of the solution, I_0 is the intensity of radiation entering and I the intensity leaving a solution of optical pathlength b cm and concentration c mol l^{-1}. Converting to logarithms and changing the signs:

$$-\log_{10} T = \log_{10} \frac{I_0}{I} \propto bc$$

$\text{Log}_{10}\ I_0/I$ is termed the *absorbance* or optical density of the solution and the relationship is called the Beer–Lambert–Bouguer Law or *Beer's Law*:

$$A = -\log_{10} T = \epsilon bc$$

where A is the absorbance of a particular wavelength and is dimensionless, and the constant of proportionality (ϵ) is termed the *molar absorptivity* at the specified wavelength and has units of M^{-1} cm^{-1} (M stands for molarity or the molecular weight in $g\,l^{-1}$). The molar absorptivity thus represents the absorbance of a 1 M solution measured in a 10 mm layer. Occasionally reference is made to the 'extinction' values of a solution expressed as $A_{1cm}^{1\%}$ or $E_{1cm}^{1\%}$, which means the absorbance of a 1% w/v solution of the solute in a 1 cm path-length. For a compound of molecular weight m, a 1% solution is $10/m$ M and so:

$$A_{1\,cm}^{1\%} = \frac{10 \times \epsilon}{m}$$

On the other hand, a few recent texts used SI units where concentration is expressed in mol m^{-3} and the molar absorptivity is for a path-length of 1 m. The SI molar absorptivity is related to ϵ by:

$$\epsilon' = 0.1\epsilon \text{ mol}^{-1} \text{ m}^2$$

Examination of the spectra given earlier in this chapter shows that the ϵ values of maxima generally fall in the range $10-10^5$ M^{-1} cm^{-1}. An absorbance of 0.3 A can be measured accurately on most instruments: a compound of $\epsilon_{max} = 10^4$ M^{-1} cm^{-1} in solution at a concentration of 30 μM measured in a 1 cm pathlength cell will have this absorbance. 1 ml of solution should suffice for this measurement, and so 30 nmol of the compound can be readily measured. By using a microcell and optimizing the conditions of measurement, the detection limit for such a compound can probably be extended into the pmol region.

1.6   Measurement of absorption spectra

The task is to measure the fraction of radiation entering the cell that is absorbed by the sample, to convert these data into absorbance values and, in the case of recording instruments, to present these values in the form of a continuous spectrum. Most instruments employ a phototube or photoconductive detector whose response is linearly related to light intensity. The instrument is arranged to measure I, the radiation intensity transmitted by the sample, and compares it to I_0, the intensity of a reference beam that does not pass through the sample. This comparison gives the transmittance, $T = I/I_0$. Conversion of T to an absorbance value was a major problem in earlier instruments since it involves a logarithmic function, i.e. $A = -\log T$, and ingeneous mechanical devices or logarithmic recorder slidewires were employed. The advent of cheap solid-state operational amplifiers for linear–log conversion, and more recently digital devices, means that the conversion is performed very accurately in modern instruments.

The fundamental problems involved in measuring the transmission are two-fold: (a) to ensure that only radiation of the specified wave-length reaches the detector, and (b) to ensure that the attenuation of the radiation is due solely to absorption by the solute. The problem associated with (a) will be discussed later in the chapter; as far as (b) is concerned, the factors causing attenuation of the measuring beam are illustrated in Fig. 1.11 and must be compensated for when measurements are made.

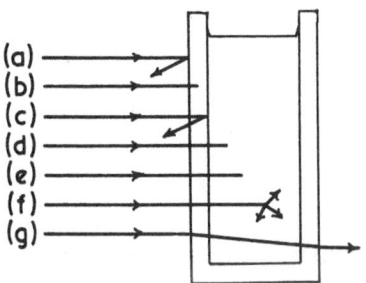

Fig. 1.11 *Possible causes of attenuation of the measuring beam as it passes through a cell containing a sample solution. (a) Reflection at air/glass interface; (b) absorption by glass; (c) reflection at glass/liquid interface; (d) absorption by solute; (e) absorption by solvent; (f) scatter by solution; (g) refraction or dispersion by cell. Accurate measurements of the solute absorbance (d) require the other factors to be minimized or compensated for.*

The earliest absorption spectra were recorded by passing 'white' radiation through the sample and into a spectrograph which dispersed the radiation by means of a prism and recorded the resulting spectrum on a photographic plate (Fig. 1.12). Beyond the problems of developing the plate, the principal difficulties were in compensating for fluctuations in the source, correcting for absorption, scatter and reflection by the cell and solvent, and calibrating the response of the plate to radiation of different wavelengths and intensities. The development of the phototube, which has a linear relationship between its

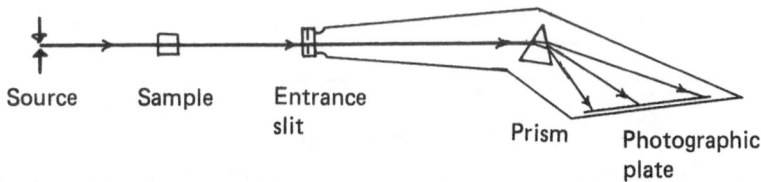

Fig. 1.12 *Schematic diagram of the arrangement for measuring absorption spectra by means of spectrograph.*

response and the light intensity, was the key to the development of new instruments for the measurement of absorption spectra. The first commercial instruments appeared in the 1940s, and since they were concerned with the measurement of both wavelength and intensity, were termed 'spectrophotometers'. Modern usage has contracted this to *spectrometer*, and the technique has become *spectrometry*. The design and construction of spectrometers will be dealt with in Chapters 2 to 6, but two properties of spectrometers that are absolutely vital to the proper measurement of absorption spectra, namely spectral bandwidth and stray-light, will be introduced here.

1.6.1   *Bandwidth*

The width of the entrance and exit slits of a monochromator determine the spread of wavelengths that emerge from it — the bandwidth of the radiation. In addition, the amount of energy passing through the instrument is also determined by the slitwidth. When measuring an absorption spectrum, the bandwidth should be as small as possible to attain the maximum spectral resolution; ideally when the monochromator is set to, say, 260 nm only radiation in a narrow spectral range centred at 260 nm should emerge from the exit slit. This is illustrated in Fig. 1.13 where (a) shows the ideal narrow rectangular

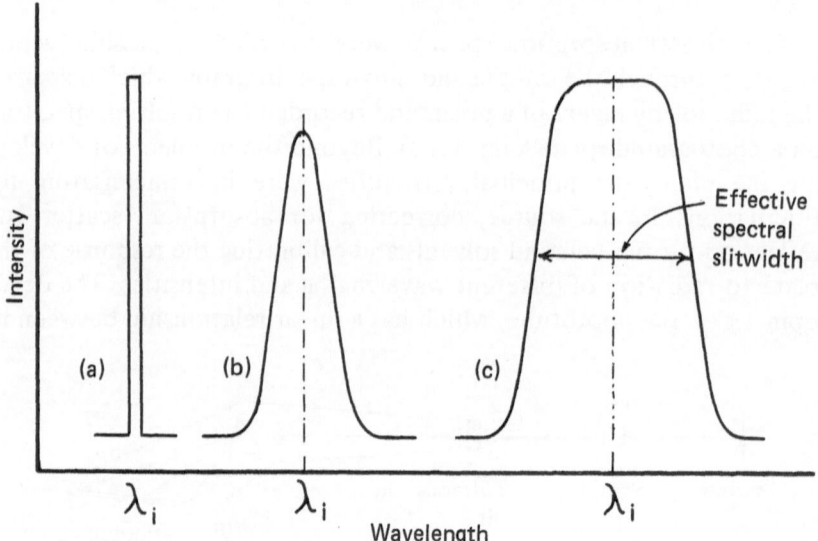

Fig. 1.13 *Spectral distribution of radiation emerging from a monochromator set at indicated wavelength λ_i. (a) Ideal profile; (b) typical profile for small slit settings; (c) profile for practical slit settings.*

profile. In reality, the spectral distribution from a monochromator with narrow slits is roughly triangular in shape, as in (b). However, even this profile cannot be used in practice for a slit opening given this profile is unlikely to pass sufficient energy for the efficient opera-tion of the instrument and so the slits must be opened further, and a typical operating profile is shown in (c). The bandwidth is best expressed as the width of the profile at half-peak height and is termed the *effective spectral slitwidth* (ESW). If the ESW approaches the width of the absorption peak to be measured, then serious degradation of the recorded spectrum will result. The true width of an absorption band at half-peak height is termed the *natural bandwidth* (NBW) and Fig. 1.14 shows how the measured height of an absorption band falls as the ratio ESW/NBW is increased. As a rule-of-thumb, the error in peak height measurement will be negligible if the ESW is kept at less than 10% of the NBW of the band.

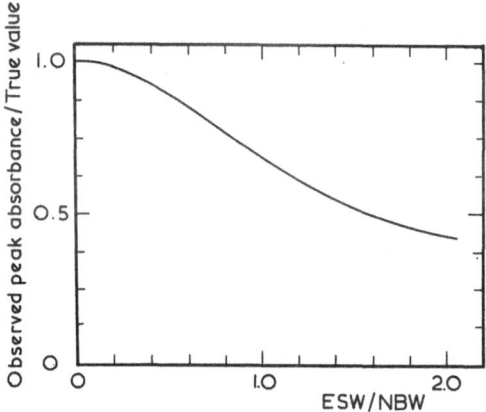

Fig. 1.14 *Effect of slit setting upon the measured height of a Gaussian-shaped absorption band.*

When compounds with closely spaced absorption maxima are measured, the peak-broadening effect due to excessive ESW will cause the individual peaks to merge into a broad band. This partially explains the differences between the benzene spectra of Figs 1.4 and 1.5. The loss of some of the peaks seen in the gas phase spectrum when the molecule was put into hexane is due in part to solvent broadening of the lines, but is also due to the greater ESW used in measuring the solution spectrum. Quantitative measurements of spectra with such narrow bands should be avoided, but if they must be made, a compromise has to be reached in selecting an ESW that is sufficiently small to resolve the bands satisfactorily while allowing sufficient energy to pass through the sample to ensure photometric accuracy.

1.6.2   *Stray-light*

When 'white' radiation from a tungsten or deuterium lamp is passed through an ideal monochromator, only the selected wavelength will emerge from the exit slit. Fig. 1.13 showed that in normal operation, a band of wavelengths a few nanometres wide is found but, in addition to this, optical defects in real monochromators causes radiation of widely differing wavelengths to emerge. Such emergent radiation of wavelengths other than the indicated wavelength is termed *stray-light*, even though it may be in the UV region. It is an inherent property of diffraction gratings that they produce several overlapping spectra of different orders, and so when set at an angle to reflect 750 nm light, the grating will also reflect radiation of 375 and 250 nm. Consequently, this effect is a major cause of stray-light in grating monochromators. Beyond this, imperfections in gratings, prisms or mirrors, reflection or refraction at slits and beam masks, and dust contamination will all contribute to the stray-light.

To demonstrate these effects, turn the wavelength setting of your spectrometer to 550 nm and open the slits as wide as possible. Place a piece of white non-fluorescent card in the sample position and note how the colour of the patch of light on the card changes from side-to-side; this is due primarily to the width of the slit and is termed *near stray-light*. Turn the wavelength down to 380 nm: the patch of light should disappear before this wavelength is reached, but you will probably still be able to see a patch of faint white *far stray-light*. This is of far lower intensity than the near stray-light but can cause far more serious distortions of spectra. Suppose that when the instrument is set at 380 nm, 1% of the radiation reaching the detector is stray-light of wavelengths greater than 400 nm. If a sample is placed in the beam which transmits only 10% of the 380 nm radiation but has effectively 100% transmission above 400 nm, then the stray-light will form 9% of the light reaching the detector. If the true absorbance of the sample at 380 nm is 1.000 A, the apparent absorbance will only be 0.963 A.

This effect becomes more significant when making measurements at shorter wavelengths for the intensity of the radiation from the lamp is less at short wavelengths and so the stray-light will be a greater proportion of the radiation reaching the detector. In addition, the sensitivity of the detector is less at shorter wavelengths and the effect of longwave stray-light will be proportionately greater. The possibility of significant errors caused by stray-light should be borne in mind when using an instrument with a tungsten lamp below 400 nm and with a deuterium lamp below 250 nm. Prism instruments are generally fitted with a UV-transmitting filter that is inserted into the beam below 400 nm, but there is no suitable filter transmitting only below 250 nm for use with a deuterium source. Grating instruments must have accessory filters to remove spectra of unwanted orders, but again restrictions on the choice of filter means that stray-light becomes a problem at shorter wavelengths.

Examples of the effects of stray-light are given in Chapter 9 and a more complete discussion of the problem appears in Volume 1 of this series [2]. Consideration of bandwidth and stray-light in mono-chromator design is discussed in Chapter 4.

References

1 Perkampus, H.H., Sandeman, I. and Timmons, C.J. (Eds) (1966), *UV Atlas of Organic Compounds*, Butterworths, London and Verlag Chemie, Weinheim.
2 Burgess, C. and Knowles, A. (Eds) (1981), *Techniques in Visible and Ultra-violet Spectrometry*, Vol. 1, *Standards in Absorption Spectrometry*, Chapman and Hall, London.

General reading

Stern, E.S. and Timmons, C.J. (1970), *Introduction to Electronic Absorption Spectroscopy in Organic Chemistry*, 3rd Edn, Arnold, London.
Steward, J.E. (Ed.) (1975), *Introduction to Ultraviolet and Visible Spectroscopy*, Pye Unicam, Cambridge.
Beaven, G.H., Johnson, E.A., Willis, H.A. and Miller, R.G.J. (1961), *Molecular Spectroscopy*, Heywood, London.
Anon. (1973), *Optimum Parameters for Spectrophotometry*, Varian Instruments, Palo Alto, California.
Edisbury, J.R. (1966), *Practical Hints on Absorption Spectrometry*, Hilger, London.
Rao, C.N.R. (1967), *Ultraviolet and Visible Spectroscopy*, 2nd Edn, Butter-worths, London.

2 Spectrometer design

2.1 Types of instrument

At the time of writing there are approximately thirty commercial organizations marketing analytical UV and/or visible spectrometers. Most of these companies sell a range of instruments embracing a wide range of performance characteristics, from a simple visible spectrometer small enough to fit into a lab-coat pocket to UV—VIS—near-IR models large enough to make a considerable hole in the pocket. This large number of spectrometers currently available, with considerable overlap in their capabilities, reflects the extent to which this area of spectroscopy is used in laboratories and the large potential market for these instruments.

In addition to the self-contained analytical instrument which usually, annoyingly, requires only the addition of a plug to the mains lead to render the instrument operational, it is possible to obtain all the appropriate components — light sources, monochromators, detectors, power supplies, optical benches, lenses, mirrors and so on — to enable the gifted amateur to build his own spectrometer. This exercise is not for the uninitiated, and should never be contemplated unless for some application for which a suitable commercial instrument does just not exist. The problems of stray-light, for example, and consequent errors in photometric accuracy are formidable and, given the current state of developments in the reduction of stray-light in instruments, it is unlikely that any home-built instrument will ever approach the performance of top-of-the-range commercial spectrometers. However, all this is not to say that for a particular application, such as monitoring a process 'on line' or measuring the gas liberated above a ploughed field, a suitable instrument cannot be constructed. In addition to the complete analytical instrument, there are a few so-called 'research' systems available. These are usually designed for a specific purpose which is not primarily analytical, such as kinetic measurements or monitoring fast reactions, and may be constructed on a modular basis so that the customer can have the instrument designed and assembled to suit the particular research area.

Because of the rapid changes in instrument design, it is not intended to provide any kind of a buyer's guide, but several publications list manufacturers' names and addresses (see, for example, [1]). A few references provide further information concerning the overall instrument, although most authors confine themselves to dealing with individual components rather than with how they go together to make a complete system. Marr's chapter in *Comprehensive Analytical Chemistry* [2] makes a good starting point, and there are several text books written by physicists, but nonetheless quite readable, such as Thorne's *Spectrophysics* [3] and Lothian's *Optics and its Uses* [4]. More detailed treatments, particularly of spectrometer design, are provided by James and Sternberg [5] and Bousquet [6]. A useful simple introduction is given in the booklet edited by Steward [7], though the content is based entirely on one manufacturer's range of instrumentation.

As indicated in the opening paragraph, the performance of current instrumentation covers a wide range. The simpler instruments will enable absorbance measurements to be made at selected wavelengths, whereas the more complex will produce a plot of absorbance versus wavelength with a high degree of accuracy, and may be able to perform operations on the resulting spectrum such as differentiation, etc. The instrument may be able to repetitively scan the same spectral region, at two wavelengths simultaneously, plot absorbance as a function of time, store or plot calibration curves, subtract a stored spectrum from another, and so on. The facilities that current instruments can offer through the in-built computing power of microprocessors make these a genuine new generation that has evolved since the publication of Edisbury's *Practical Hints* in 1966 [8]. There have also been developments in the non-computing hardware, such as the use of holographic gratings, silica-coated optics, pulsed light sources and so on, which have considerably improved instrument performance since the mid-sixties, not only at the high absorbance end of the scale but also at the low absorbance end as the demand for sensitive detectors for HPLC continues to increase (see, for example, [9]).

A range of accessories are now also available for measuring fluorescence or diffuse reflectance, for scanning TLC or gel chromatography plates, for measuring colour and opaque samples by means of an external integrating sphere, and so on. Some of these accessories, such as that for measuring fluorescence, will not produce as good a performance as a purpose-built instrument, but if a good UV–VIS instrument is available, the accessory will provide a low-cost alternative to the purchase of a fluorimeter if only a small number of measurements are to be made. The usefulness of accessories has been evaluated by Clark [10] and Mills [11]. Manufacturers will be only too happy to supply details of the accessories available for their product range.

2.2   Basic arrangement of optical components

Despite the large number of different instruments available, which must number nearly one hundred, the arrangement of the optical components within the spectrometer in the majority of these is very similar. In general terms the spectrometer will consist of the following.

(a)   *A light source or sources*

For reasons which will be discussed in Chapter 3, it is not possible to cover the range from 200 nm through the visible to the onset of the IR at 800 nm with a single light source and so an instrument which covers the whole of this spectral range will require two sources. Sources will be dealt with in Chapter 3.

(b)   *A device for spatially separating the light from the source into its component wavelengths and isolating a portion of this spectrum*

In the vast majority of instruments this function is performed by a monochromator containing a prism or diffraction grating as the dispersing element together with entrance and exit slits; monochromators will be discussed in detail in Chapter 4. Some instruments may contain two monochromators arranged to give increased dispersion and also lower stray-light levels or the monochromator may contain two dispersing devices such as the Echelle monochromator [12].

At the other end of the performance scale the wavelength separation may be performed by coloured glass filters or an interference filter wedge. This method of wavelength separation, although inexpensive compared with the use of a prism or grating monochromator, suffers from two distinct disadvantages. Firstly the range of wavelengths transmitted at any given setting (the passband) can be quite large: as much as several tens of nanometres. Such a large passband may cause deviations from Beer's law. Secondly the transmission of such devices may be quite low and as much as 90% of the incident radiation may be lost with filters in the 200–250 nm range. The use of filters for wavelength selection will be discussed further in Section 4.9.

(c) *A sample compartment or compartments so that the sample material – usually a solution in a suitable cell – may interact with the light beam*

In addition the sample compartment may be able to accommodate cells for gaseous samples or an accessory for the measurement of specular reflection from solid samples. There may also be a thermostat device to maintain the sample at a constant temperature or a means of changing the sample solution automatically. The instrument may have a 'turbid sample' or 'scattered transmission' arrangement consisting of an extra sample position located close to the detector. Cell compartments and sample handling will be dealt with in Chapters 8 and 9.

(d) *A detector to measure the light intensity before and after interaction with the sample*

In most instruments this will be the electron multiplier phototube, known by its more common but incorrect designation of 'photomultiplier' tube. Less expensive instruments may have a single-stage photo-emissive tube and instruments responding in the near-IR will have a photoconductive device like the lead sulphide cell. Recently, diode arrays have become available as detectors for spectrometers enabling the complete spectrum to be recorded in a few milliseconds. All wavelengths of the spectrum fall on the diode array simultaneously and each diode is discharged to an extent dependent on the light intensity at a particular wavelength. The array is processed by a sequential recharging current.

All these devices are photon detectors as opposed to thermal detectors (thermocouple, bolometer, pyroelectric radiometer, etc.) and the arrival of photons causes a charge carrier – usually an electron – to be raised to a higher energy state. The response of all detectors depends on wavelength, being a function of the properties of the photosensitive element and the window transmission characteristics, and it may be necessary to have more than one device to cover the whole spectral range. If an instrument is to be used over a spectral region for which it was not originally designed, the detector may have to be changed. Detectors are discussed in detail in Chapter 5.

(e)   *Optical components including lenses, mirrors, beam-splitters, choppers and windows*

The function of these components is to get as much light as possible from the source to the dispersing device, then to the sample and then to the detector. In addition, the dimensions of the beam may have to be altered, for example, so that it all passes through a small area in the sample compartment, and it may need to be focused at particular locations, such as on the entrance slit of the monochromator. Also the need to produce a compact instrument may necessitate bending the light beam round a few corners by means of mirrors: as mirror surfaces never have reflectances of 100%, it is important that the number of bends be kept to a minimum. In double-beam instruments, the light beam is split into two components one of which passes through the sample and the other through a reference cell: these instruments must also have a beam-splitter device to achieve this, and possibly also a further device to recombine the beams. A single-beam instrument may be improved by the inclusion of an optical chopper to convert the output of a source to a.c. in order that an a.c. detector and amplifier circuit may be used.

(f)   *Electronics, electrical and mechanical devices*

There will be stabilized power supplies to run the sources and detectors. The detector signal will need to be amplified and processed to give a readout of the sample absorbance. The lamp changeover mechanism will need to be activated and driven, as will the wavelength drive of the monochromator, and possibly also the filters used to remove stray-light and unwanted spectral orders from the grating. The slitwidth control and sample holder may also be motor driven. The instrument may incorporate a microprocessor both to operate certain functions and to process and store results. There may be some safety interlocks such as that to prevent a photomultiplier being overloaded with room light on opening the sample compartment door.

(g)   *A box to put the whole lot in*

This provides mechanical stability for the optical components, a light-tight housing and some protection from chemical attack. It needs to be designed so that certain components are readily accessible for (i) changing in the event of failure such as the lamps or detectors and (ii) cleaning, as optical surfaces inevitably become a resting place for dust, though the cover should discourage tinkering with the mono-chromator. Adjustment of the internal components of the mono-chromator should only be contemplated as a last resort for this should not be necessary with a properly maintained instrument, though it may be necessary to clean the components from time to time. It is worth mentioning that unless the optical surfaces have been silica coated, as in some recent instruments, they are extremely susceptible to mechanical damage by abrasion during the cleaning process (details in Chapter 13). A surgical-cleanliness should be observed when any part of the instrument is exposed which is not normally open to the laboratory atmosphere. The maxim of 'leave well alone' is strongly recommended.

2.3 Single- and double-beam instruments

In the simplest spectrometer, a single beam of light is taken from the source and travels through the various optical components of the detector. A typical single-beam spectrometer is shown in Fig. 2.1. In order to measure a spectrum with such an instrument, at each wave-length the absorbance scale must first be reset to zero with a reference cell in the light beam before the sample absorbance can be read. This procedure is necessary because the output of the source, the detector response and the transmission and reflectivity of the various optical components all vary with wavelength.

In double-beam spectrometers, a second beam of radiation from the source follows a similar path through the instrument but passes through the reference cell. The difference in intensity between the reference and sample beams thus represents the light absorbed by the sample since the other variables in the system are compensated for.

The light beam is split into two beams of equal intensity after the radiation has been dispersed. This is normally done by a rotating sector mirror which produces two square-wave wave-trains, 180° out of phase with each other. One of these beams passes through the sample and the other through the reference cell, usually located side-by-side in the sample compartment, though not too close together, otherwise the sample and reference beam overlap and 'cross talk' results.

After passing through the sample compartment the beams follow

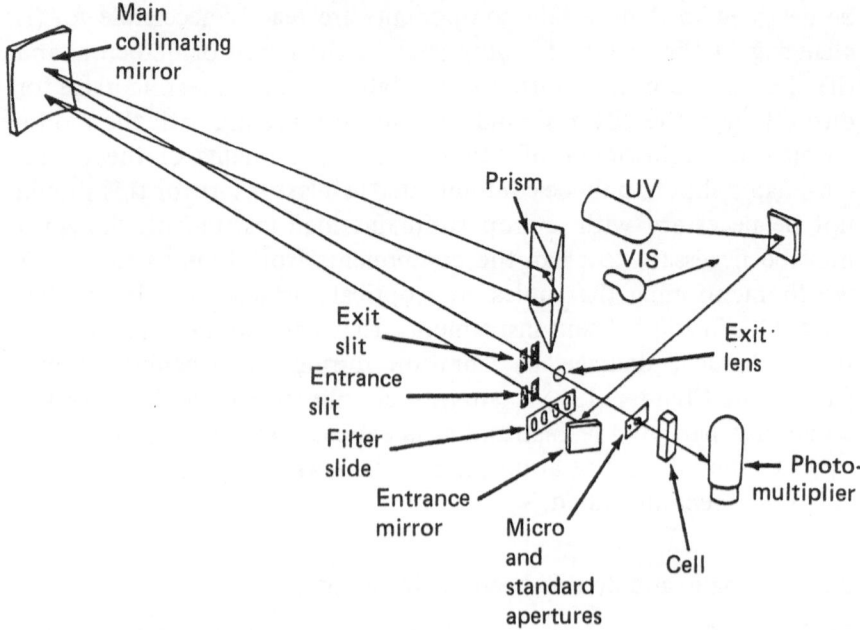

Fig. 2.1 *Optical arrangement of a typical single-beam spectrometer. This instrument has two sources selected by a mirror, and a prism monochromator. Reproduced by courtesy of Gilford Instruments.*

independent optical paths to the detector. Current practice is to use a single detector for both beams and to provide a reference signal from the chopper by a simple LED source and suitable detector so that the instrument electronics can recognize which output from the detector is due to the sample beam and to the reference beam. This arrangement is normally referred to as 'double-beam in time'. The alternative arrangement, in which two detectors are used, one for each beam, is referred to as 'double-beam in space'. This latter arrangement has been used in commercial instruments, but is not favoured by manufacturers at the present. A diagram of the appropriate part of a double-beam instrument is shown in Fig. 2.2. In this instrument a single rotating sector mirror is used, but there are instruments in which a second sector alternately directs the beams onto the detector as shown in Fig. 2.3. Not all instruments use a rotating sector to achieve the double-beam capability and an instrument which uses a beam-splitter to divert half the incident beam through the sample and half through the reference simultaneously is shown in Fig. 2.4. The beam-splitter is made from two plane front-surfaced

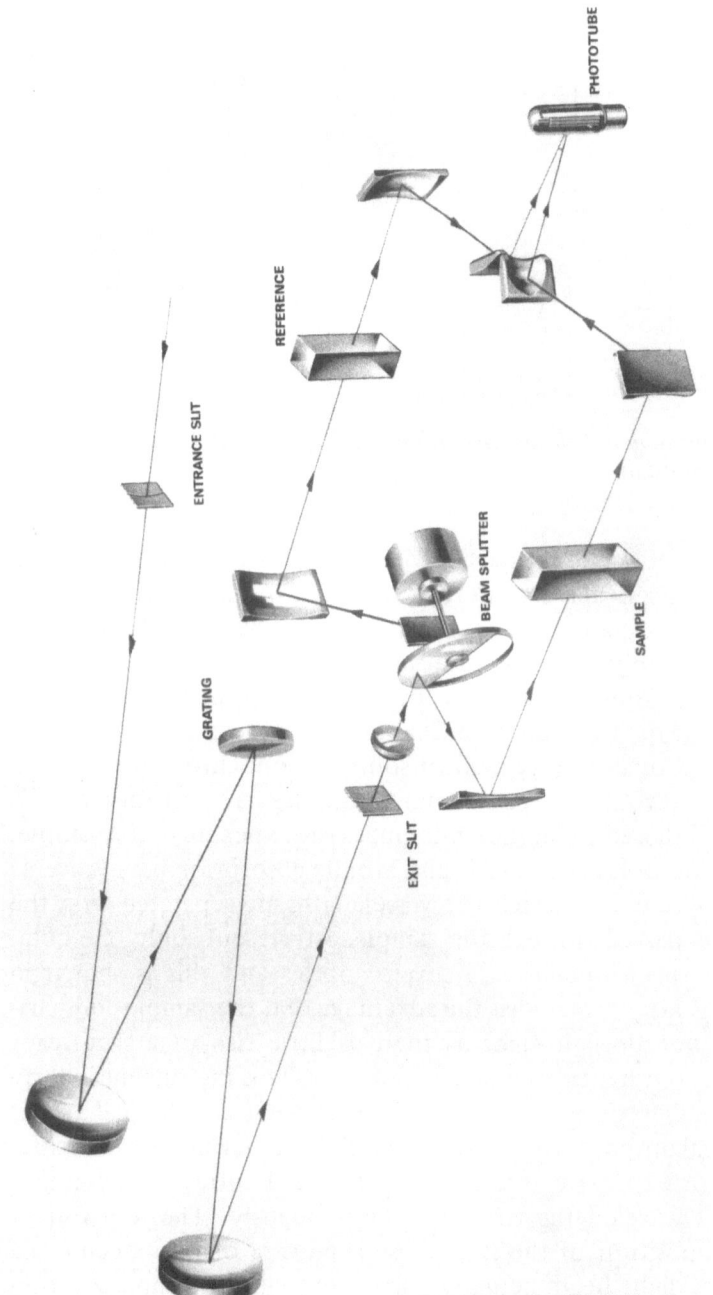

Fig. 2.2 *A typical double-beam spectrometer using a rotating sector mirror to generate the two beams. Reproduced by permission of Varian Associated Ltd.*

PHOTOTUBE

REFERENCE

ENTRANCE SLIT

BEAM SPLITTER

SAMPLE

GRATING

EXIT SLIT

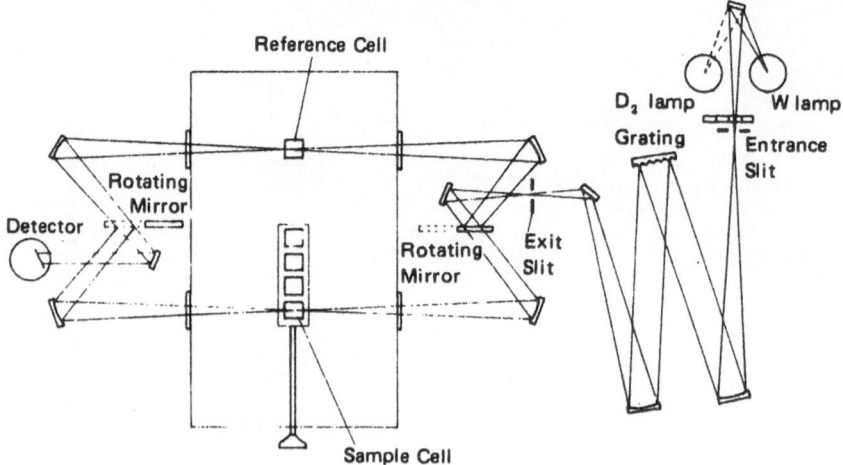

Fig. 2.3 *A typical double-beam spectrometer. Reproduced by permission of Shimadzu Corporation.*

mirrors set at an angle, rather like the roof of a house; the device is thus sometimes called a 'roof mirror'. In order to distinguish between the beams a rotating chopper is used. In most instruments, whether using the beam-splitter or the rotating sector mirror, the light is completely cut off from the detector for part of the time so that a measurement of the dark current (see Chapter 6) can be made.

The conventional arrangement of source—monochromator—sample —detector is favoured because many samples are sensitive to UV radiation and this configuration minimizes the exposure of the sample. Other spectroscopies such as IR and atomic absorption use a 'reverse optics' arrangement in which the wavelengths are separated *after* the radiation has passed through the sample. At present, only one commercial UV spectrometer uses reverse optics and this is shown in Fig. 2.5. This arrangement has the advantage that the sample compartment need not be light-tight as there is little risk of a significant amount of room light reaching the detector. This instrument incorporates a number of other unique features such as elliptical rather than spherical optics — which have better light gathering characteristics but are more expensive to make — and a diode array detector that measures all wavelengths virtually simultaneously. The problem of photodecomposition of the sample is taken care of by reducing the time that the light beam actually passes through the sample solution to about 200 ms.

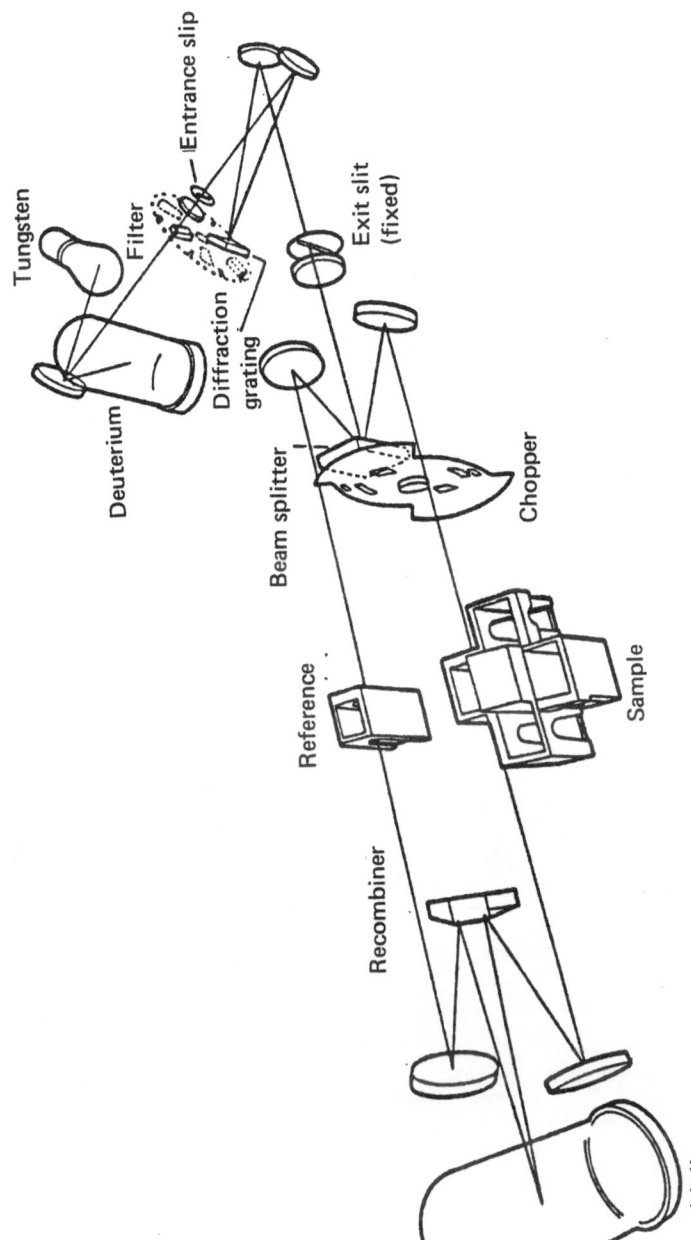

Tungsten

Filter

Entrance slip

Deuterium

Diffraction grating

Exit slit (fixed)

Beam splitter

Chopper

Reference

Sample

Recombiner

Photomultiplier

Fig. 2.4 A typical double-beam spectrometer using a beam-splitter mirror. A second mirror device is used to direct the two beams onto the same region of the photomultiplier cathode. Reproduced by courtesy of Bausch and Lomb.

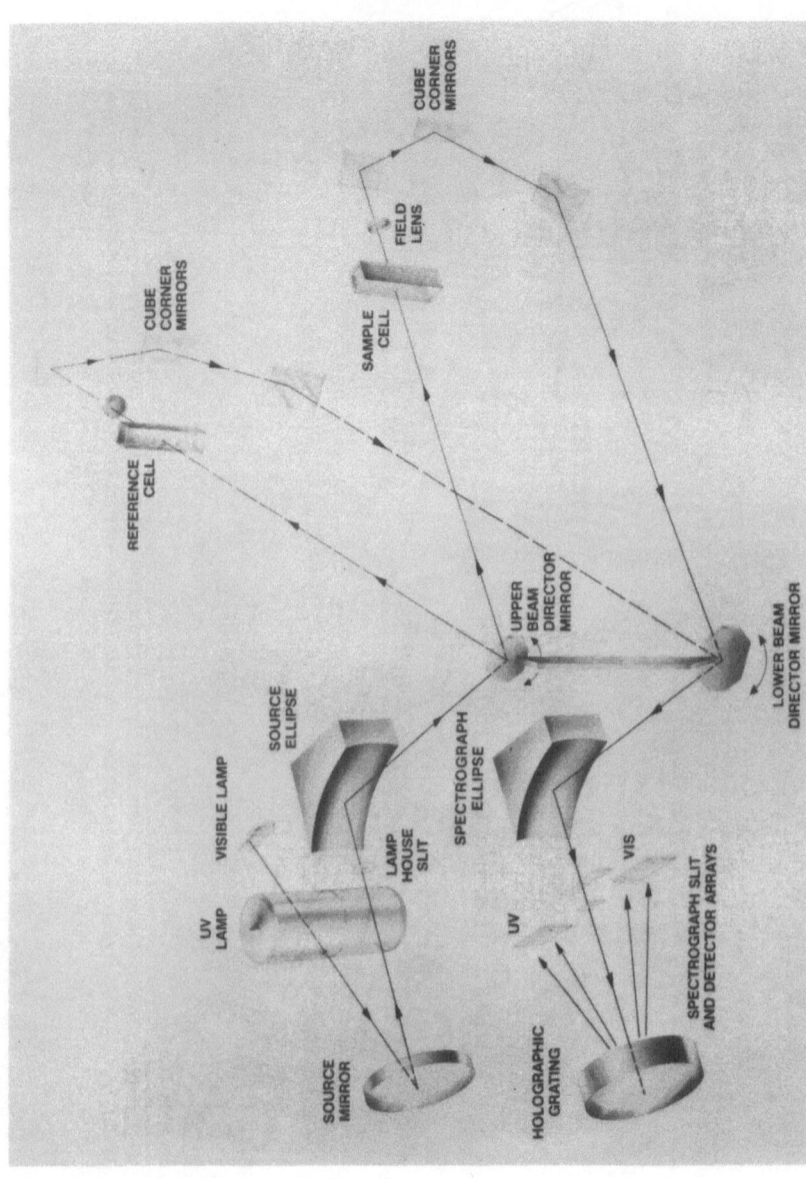

Fig. 2.5 A 'reverse-optics' spectrometer using diode array detectors. Reproduced from [13] by courtesy of Hewlett–Packard Ltd.

One of the major features which affects photometric accuracy is the relative intensity of the stray-light component compared to the primary wavelength component to the measuring beam, the problem being particularly acute at high absorbance values or at wavelengths at which the source provides low intensities [14]. Thus for the highest photometric accuracy and adherence to a linear absorbance—concentration relationship at high absorbances, it is important that as much as possible of the radiation from the source passes through the sample to the detector. The additional reflecting surfaces of the double-beam system cause a loss in energy compared with the single-beam system. This loss will be worse at shorter wavelengths and will increase as the optical components age. As the sample and reference beams follow different paths through the instrument, any mismatch in the optical components, cell positioning, windows, etc., between the two paths will cause inaccuracies in measurement. In addition there will be problems of mechanical stability due to the incorporation of a high-speed rotating optical component — the chopping frequency is typically between 50 and 100 Hz — which may cause vibration of the other components. Although the double-beam system will compensate for slow drift in the source and photometer output, it will not in any way compensate for noise arising from the electronic components of the lamp power supply or, more important, the detector circuitry. This noise will thus exert a greater effect in the double-beam system because of the reduced intensity in each optical path due to the additional optical components. This effect will be worse for those instruments which use a beam-splitter rather than a rotating mirror to achieve the double beam, as in the former only a nominal 50% of the light from the source passes through each beam, whereas in the latter a nominal 100% is directed alternately into each beam.

Until recently, instrument development has been along the lines of increasing the photometric accuracy of double-beam spectrometers and so the 'top of the range' of the present generation of instruments can give single-beam photometric accuracy. Considerable effort has been devoted to reducing stray-light, noise, increasing energy throughput, etc. There has been one exception to this approach in that Gilford Instruments produced a single-beam instrument which achieved the double-beam spectrum recording capability by rapidly moving the sample and reference cells in and out of the light beam at about five times per second. Another approach with microprocessor instruments having suitable memory is to measure the reference spectrum over the required wavelength range and then store it in the computer's memory. This spectrum can then be subtracted from the sample spectrum obtained over the same wavelength range. The sample and reference solutions need only be moved once.

Such instruments are just now becoming commercially available. For example, in the Beckman and Gilford ranges the instruments achieve the 'double-beam' wavelength scanning ability in the manner described above (together with computer control over cell repositioning), though the manufacturers have had to pay particular attention to source and detector stability as the sample and reference spectra are not obtained at the same time but sequentially. This also requires accuracy and precision of wavelength setting. The electronics of the detector circuitry can give very fast responses so that a measurement can be made every 50 ms and thus scan rates as high as 1200 nm min^{-1} can be achieved.

In conclusion, it can be seen that a double-beam instrument is more complex than a single-beam and it is to be expected that if two instruments have similar performances in terms of monochromators, throughput, stray-light reduction, etc., the double-beam instrument will be more expensive than the single-beam variety. The relative complexities of these two configurations is only one criterion to be taken into account when choosing an instrument for a particular job. The following chapters will provide guidance on such criteria as resolution, stray-light level, baseline flatness, noise, data handling, accessories, etc. All of these criteria will need to be balanced against the various costs involved, capital outlay and maintenance being possibly the most important. The availability of service engineers and spares should also be borne in mind in these days of readily available imported instrumentation.

References

1 *International Laboratory Buyers Guide*, International Scientific Communications Inc., 808 Kings Highway, Fairfield, Connecticut 06430, USA. Published annually.
2 Marr, I.L. (1975), in *Comprehensive Analytical Chemistry* Vol. IV (Ed C.L. Wilson), Elsevier, Amsterdam.
3 Thorne, A.P. (1974), *Spectrophysics*, Chapman and Hall, London.
4 Lothian, G.F. (1975), *Optics and its Uses*, Van Nostrand Reinhold, London.
5 James, J.F. and Sternberg, R.S. (1969), *Design of Optical Spectrometers*, Chapman and Hall, London.
6 Bousquet, P. (1971), *Spectroscopy and its Instrumentation*, Hilger, London.

7 Steward, J.E. (Ed) (1975), *Introduction to Ultraviolet and Visible Spectrophotometry*, Pye Unicam, Cambridge.
8 Edisbury, J.R. (1966), *Practical Hints on Absorption Spectroscopy*, Hilger and Watts, London.
9 Various authors (1977), *UV Spectrom. Grp Bull.*, 5 supplement, 2–42.
10 Clark, P.A. (1976), *UV Spectrom. Grp Bull.*, 4, 6.
11 Mills, K.J. (1976), *UV Spectrom. Grp Bull.*, 4, 11.
12 Keliher, P.N. and Wohlers, C.C. (1976), *Anal. Chem.*, 48, 333A.
13 Anon. (1980), *Hewlett Packard Journal*, February.
14 Francis, R.J. (1980), *International Laboratory*, May/June, 85.

3 Light sources and optical components

In general terms, the function of the light source is to produce radiation of the appropriate wavelength and of sufficient intensity to allow a measurement after interaction with the sample material. An ideal source should have spatial and temporal stability, and uniform power output per wavelength interval across the spectral region of interest. The sources that are commonly used in UV–VIS spectrometers emit radiation continuously and are intrinsically stable, although, of course, freedom from noise and drift also demands a stable power supply. No source produces a uniform power output at all wavelengths and no one source adequately covers the UV and visible region of the spectrum. An instrument covering the whole spectral range must have two sources that are usually interchanged in the optical path by the movement of a mirror but sometimes by moving the lamps themselves. As will be seen later, the two sources will have different physical sizes and the changeover can cause problems of optical alignment and light gathering. The optics of the instrument may be designed, for example, to produce an image of the UV source which just fills the entrance slit of the monochromator thus using the monochromator to its greatest effect, but it may not be possible to do this with the light from the visible source as well.

Sources have a limited lifetime and will need to be replaced periodically. A deterioration in the signal-to-noise characteristics will usually be noticed before catastrophic failure occurs. For UV–VIS spectrometers the sources used can be conveniently classified as (a) gas discharges and (b) thermal radiators. Classification according to spectral region is somewhat arbitrary as most sources emit some radiation over a greater range of wavelengths than is encompassed by the terms UV, visible or IR.

The optical system of the instrument normally consists of a sequence of reflective surfaces together with the occasional lens.

Lenses suffer from a number of aberrations which are difficult and costly to correct for, so mirrors are the preferred means of directing the light along the optical path. Recently instruments using fibre optic light guides have appeared on the market [1].

3.1 Sources for the UV region

3.1.1 *Deuterium arc*

In absorption spectrometers the most commonly used source is the deuterium arc. This has a three- to five-fold increase in output over the hydrogen arc that was used formerly. Before an arc can be struck between the electrodes, the cathode has to be heated to a temperature of a few hundred degrees by means of a low voltage filament. A starting voltage of several hundred volts then initiates the arc which produces enough heat to maintain the discharge and the heating current can be switched off. During normal running, the voltage is 80−100 V and the current flow is a few hundred milliamps giving a power consumption of between 30 and 60 W. The arc is confined to a small volume by a metal enclosure around the electrodes and the radiation escapes through a small circular (1 mm diameter) or rectangular (1 mm x 5 mm) aperture. A typical lifetime is 500 h and the lamp should be switched off when not in use to conserve its life: a wise practice in view of the high cost of replacement. The spectral output is a continuum from 180 to 400 nm with a few scattered lines in the visible which are useful for wavelength checking purposes (e.g. lines at 486.0 and 656.1 nm), and thus the lamp envelope must be capable of transmitting this range of wavelengths: usually the lamp is made from a suitable grade of silica. A typical lamp is shown in Fig. 3.1a and the spectral output as a function of wavelength is shown in Fig. 3.3.

3.1.2 *Mercury and xenon arcs*

For some absorbance applications and for other types of analytical spectroscopy, such as luminescence or photoacoustic, a deuterium arc source is not sufficiently intense and a higher power mercury or xenon arc is used. Lamps can be obtained with powers up to 2000 W. Both mercury and xenon lamps are essentially similar in construction consisting of two tungsten electrodes spaced 1−8 mm apart and sealed inside a quartz envelope. The gas filling is at such a pressure that under normal - operating conditions a pressure of 750−1200 psi

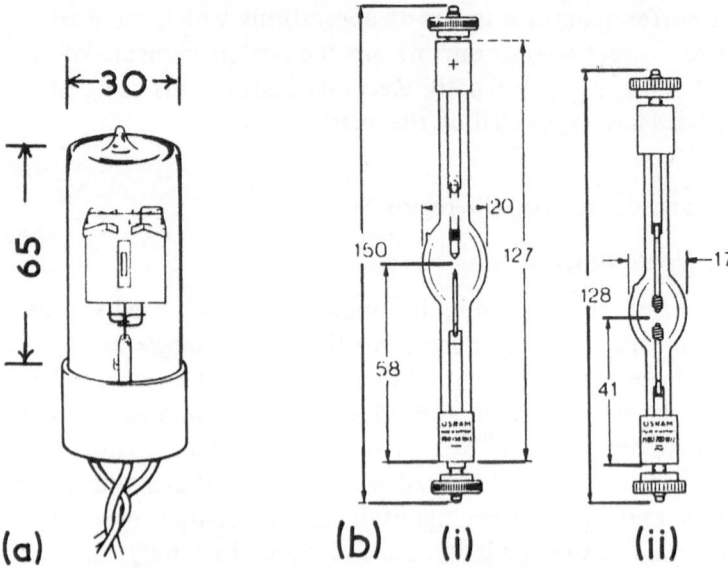

Fig. 3.1 *(a) A typical deuterium arc lamp. Dimensions in mm. Reproduced by permission of Pye Unicam Ltd. (b) Typical high-pressure arc lamps: (i) 150 W xenon lamp (ii) 200 W mercury lamp. Dimensions in mm. Reproduced by permission of Pye Unicam Ltd.*

develops. Both lamps produce a severely broadened (by Doppler and pressure broadening) atomic line spectrum, as well as a plasma recombination continuum. The xenon lamp emits a continuous spectrum between 190 and 750 nm, some intense lines between 780 and 1000 nm and then a decreasing continuum to 2.6 μm. The mercury lamp contains a small amount of argon as starting gas and a pool of liquid mercury when cold, and emits a series of mercury lines between 240 and 700 nm and a continuum in the IR up to 2.6 μm. Thus both lamps act as sources for the near-IR. Mercury lamps are available that contain a small amount of xenon to improve stability and warm-up characteristics.

The power supply needs to provide a high-voltage pulse (10−25 kV) to initiate the discharge followed by a medium voltage to establish the arc (70−120 V), then finally the operating voltage. Xenon lamps operate at about 20 V, mercury lamps start at 10−15 V and then rise to 35−80 V as they heat up. The lifetime is also dependent on the type of lamp and varies from 200 to 2000 h. Mercury lamps take about 15 min to reach thermal stability. Typical mercury and xenon lamps are also shown in Fig. 3.1b and spectral output curves in Fig. 3.3.

3.1.3   *Hazards of UV lamps and radiation*

Mercury and xenon arc lamps run at high pressures and are above atmospheric pressure even when cold, and are thus subject to explosion. Obviously they must be operated inside a protective housing and handled with protective gloves; protective goggles should also be worn. Exposure of skin to UV radiation from any source causes erythema (reddening) or, in severe cases, blistering with secondary effects. Usually the symptoms subside after a few days with no lasting effects but, it has been shown that radiation below 320 nm is carcinogenic to some animals and there are many cases of skin cancer reported in people exposed for long periods to solar radiation. UV radiation also causes keratoconjunctivitis (inflammation of the cornea and conjunctiva) which can be irritating and painful, though again the symptoms normally subside after a few days without permanent damage. Consequently, the radiation from UV lamps should never be viewed directly unless protective goggles are worn. The permissible exposure levels recommended by the American Conference of Governmental Industrial Hygienists (ACGIH) are given in a recent article [2].

In addition, radiation below 250 nm will dissociate molecular oxygen with the formation of ozone, which has a threshold limit value set by ACGIH of 0.1 ppm [3]. Consequently, low-wattage lamps (all deuterium, xenon below 200 W and mercury below 500 W) should be operated in a well-ventilated area, and larger lamps should be vented out of the room. Ozone-free lamps are available with envelopes which begin to absorb at 300 nm and are opaque below 240 nm.

3.2   Sources for the visible region

3.2.1   *Tungsten filament bulbs*

The most commonly used source of visible radiation is an incandescent tungsten filament. In its simplest form, this may be a car headlamp bulb with a factory-mounted base so that the lamp can only be mounted in one orientation (Fig. 3.2a). The spectral distribution of energy from the filament approximates to that from the ideal incandescent body — a 'black body' — and a tungsten filament bulb usually operates at about 2900 K. The maximum intensity is around 1000 nm and falls rapidly at shorter wavelengths with the lower limit of useful output being about 330 nm. The emission also extends well into the

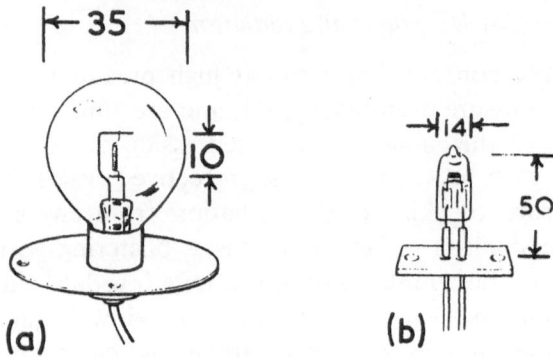

Fig. 3.2 *(a) A typical tungsten filament lamp with pre-focus base fitted. Repro-duced by permission of Pye Unicam Ltd. Dimensions in mm. (b) A typical tungsten–halogen lamp mounted in a special holder. Dimensions in mm. Repro-duced by permission of Pye Unicam Ltd.*

near-IR and the source can be used as an IR source, provided the envelope has suitable transmission characteristics.

More recently, it has become common practice to use a variant of this source known as the tungsten–halogen lamp or quartz–iodine lamp. The quartz envelope of these lamps transmits further into the UV region, and their UV output is also increased as the filament temperature is about 3500 K. They also have considerably improved long-term properties as the gradual blackening of the envelope by tungsten evaporated from the filament is reduced by reaction with the halogen and the cycle is completed by the decomposition of the tungsten halide formed by contact with the hot filament. As with ordinary tungsten bulbs, the lifetime is prolonged if they are run at a reduced voltage although care must be taken as a reduction by more than 10% may drastically shorten the lamp life due to disrup-tion of the halogen cycle. Tungsten–halogen bulbs are available at powers up to 1000 W and require a stabilized power supply capable of supplying between 8 and 13 V at high current. At 450 nm the output varies with the 8th power of the current, so the requirements of the power supply are quite stringent. A typical bulb is shown in Fig. 3.2b and the spectral output is shown in Fig. 3.3 for a 100 W lamp.

3.3   New sources

3.3.1   *Lasers*

Compared with incandescent filaments and discharges through gases, lasers are comparatively new sources of radiation. The special proper-

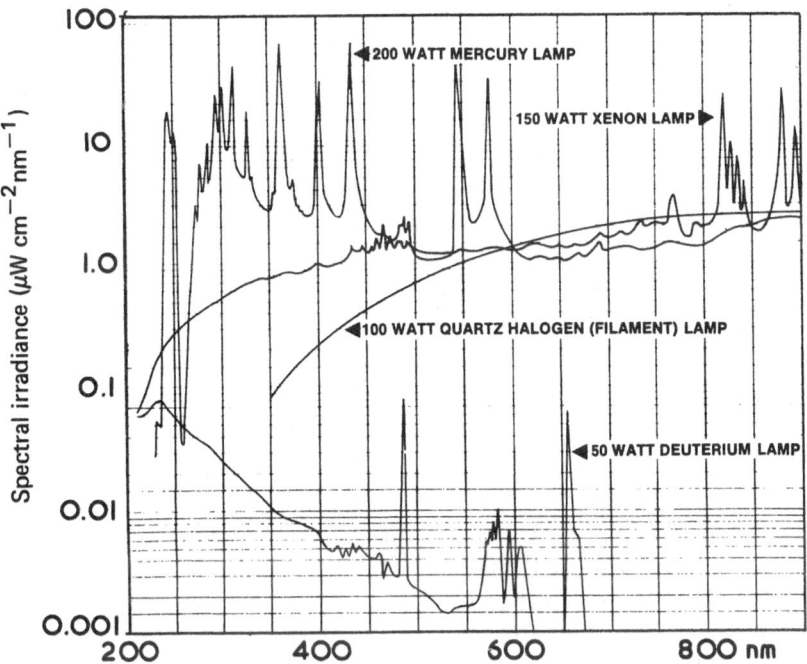

Fig. 3.3 *Spectral distribution of the outputs from typical deuterium, xenon and mercury arc lamps and a tungsten–halogen filament lamp. There were measured as the irradiance at a surface 500 mm from the lamp and plotted on a logarithmic scale. Reproduced by permission of Oriel Corporation.*

ties of a laser, which include spatial and temporal coherence, very narrow bandwidth, and high power outputs – though usually only for a short time – has made this source extremely useful in a wide range of spectroscopic techniques, particularly in the field of photo-chemistry. The developments in laser sources have given considerable impetus to a number of techniques such as coherent anti-Stokes Raman scattering and coherent Stokes Raman scattering [4] or optogalvanic spectroscopy [5], and has extended the range of applications of techniques such as photoacoustic and fluorescence spectroscopy. Explanations of the basic function of laser action and the various ways in which lasers may be operated are readily available in the literature (see, for example, [6]) and the use of the laser in analytical spectroscopy has recently been briefly reviewed [7]. The extent of the developments since Maiman's original report [8] of the ruby laser in 1960 can be seen from the size of West's review chapters in the *Periodical Reports on Photochemistry*: in Volume 10 for example, in a review covering the period July 1976 to June 1978, no less than 400 references are cited for developments in laser sources [9]. Incidentally, these chapters also provide a succinct review of UV—VIS absorption spectrometry, as well as other analytical spectrometries. Of particular interest to UV—VIS spectroscopists are the reports of broadly-tunable UV continuous-wave lasers, such as one covering the range 285—400 nm based on a dye laser [10].

At present there are relatively few types of laser commercially available and these are expensive, so that their use as sources in analytical UV—VIS spectrometers is still some way off. It seems unlikely that laser sources will ever replace the sources discussed earlier, as the laser cannot really offer any substantial advantages for conventional solution spectrometry although some applications may make use of the high peak power of the laser to measure a highly absorbing sample, or the high degree of collimation of the laser beam in measurements over a very long pathlength. However, this is not to say that lasers have nothing to offer UV—VIS spectrometry for at least two new potential analytical methods are under investigation. In the first of these, 'intracavity absorption', the sample is placed in a highly reflective, low-loss resonator cavity [11]. As absorption by the sample attenuates the light, the losses resulting from the succession of many passes back and forth in the laser cavity means that the laser output is very sensitive to the sample absorbance. In the second technique, 'thermal lensing', a highly directional laser beam is passed through an absorbing sample causing a small temperature rise that generates a refractive index gradient in the material [12]. This acts as a lens to defocus the laser beam and so the emergent beam is broader than if the sample were non-absorbing. The effect may be monitored by either measuring the change in beam profile or the power at a particular location in the profile.

3.3.2   *Pulsed sources*

In most UV—VIS spectrometers the light source is pulsed, as far as the detector and associated electronics are concerned, because of optical choppers. This modulation is useful since it allows the instrument to discriminate against constant intensity stray-light such as room light leaking into the sample compartment. Optical chopping reduces the intensity in each optical path and hence sacrifices something in terms of photometric accuracy. Recently, however, new sources have become commercially available for spectrometers using pulsed power supplies to achieve very high intensities with a low average power consumption.

The spectral output from a pulsed low-pressure xenon lamp is shown in Fig. 3.4. Comparison with the xenon curve in Fig. 3.3 shows that this source has a considerably improved UV output. The peak power may be as high as 40 kW but since the pulse width is about 1 μs, the average power consumption can be less than 5 W. This means that there is no excess heat to be removed and, perhaps more important, no ozone removal problems which can cause difficulties with the siting of instruments. These sources are compact and may have a cost advantage over existing sources. A high-voltage, high-current density discharge in a nitrogen atmosphere can produce an emission as short as 6 μs with a peak power of around 20 kW. The spectral distribution is that of a black body emitter at a temperature of 10 000 K. The intensity maximum is thus below 300 nm and the lamp has a useful spectral output from 200 to 3500 nm.

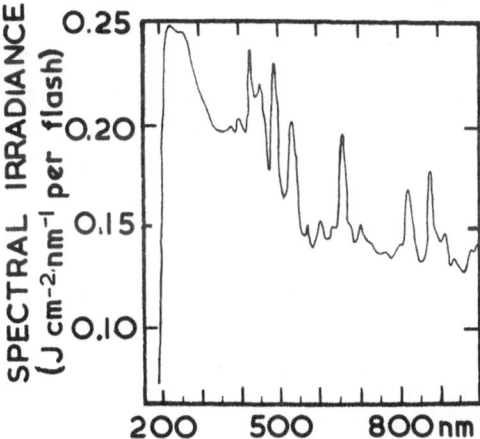

Fig. 3.4 *Spectral distribution of the output from a typical pulsed xenon lamp. The emission was measured as the peak power falling on a surface at 76 mm from the lamp. The pulse width was 6.5 μs and the total emitted power per flash was about 40 kW. The lower xenon pressure results in a relatively high UV output compared to the continuous xenon lamp shown in Fig. 3.3.*

3.4   Lenses and mirrors

Most UV–VIS spectrometers use mirrors rather than lenses, and although the mirrors must have resistance to mechanical damage and chemical attack, they are front-surfaced rather than solid metal. There are two reasons for this: mechanical stability is a problem with solid metal mirrors, and the reflectivity obtained by polishing is poorer than for a vacuum-deposited film. It is the reflectivity and how it changes with time that is of over-riding concern to the manufacturer, and should be so to the operator as well. The reason for this is not hard to see when it is realized that an instrument may contain as many as 12 reflecting surfaces between source and detector. The fraction of light getting through an optical system as a function of the number of reflections is shown in Fig. 3.5 for a number of reflectivities. The best reflectivity that can be obtained is about 90% at 200 nm for a brand new mirror produced by thermal evaporation of aluminium, which is the only metal of practical importance. Reflectivity is a function of wavelength and mirrors which differ almost by a factor of two in this respect for UV radiation may appear identical for visible light. The reason for this is that the light loss on reflection is due to two factors, namely (i) scatter from surface dust, which can exhibit a dependence on wavelength as severe as proportional to the inverse fourth power, and (ii) absorption by an alumina surface layer, which is greater in the UV. The change in reflectivity with time is shown in Fig. 3.6 from which it can be seen that the conditions in the lamp housing cause severe degradation of the reflection properties compared with a mirror exposed to air. Reference to Fig. 3.5 shows the drastic effect this degradation can have on the amount of light eventually arriving at the detector, even for an instrument with a modest number of reflections, such as 5 or 6.

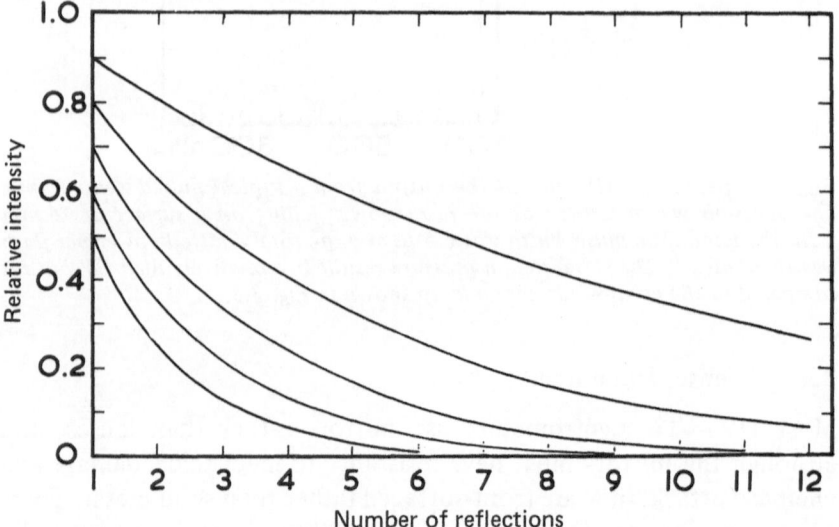

Fig. 3.5 *The attenuation of light by multiple reflections from imperfect mirrors. The curves show the fraction of light after a given number of reflections from mirrors of 90%, 80%, 70%, 60% and 50% reflectivity.*

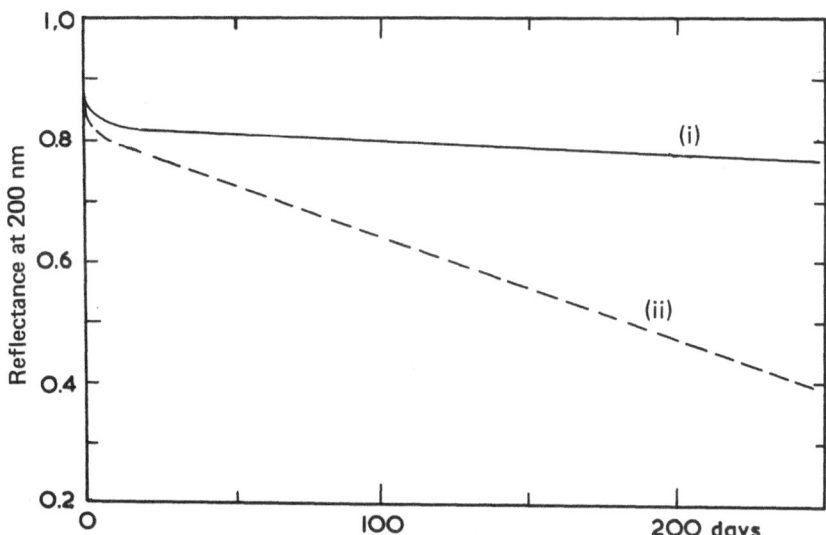

Fig. 3.6 *The effect of ageing on the reflectance at 200 nm of an uncoated aluminium mirror: (i) mirror kept in air; (ii) mirror continuously exposed to UV radiation in a spectrometer lamp-house. From [14].*

In turn, of course, the photometric accuracy degrades as the proportion of stray-light increases. One approach to reducing this problem has been to coat the aluminium surface with a thin layer of a material transparent over the required wavelength range in order to stabilize the reflectance. Magnesium fluoride has been used for this purpose but provides a relatively soft coating with little resistance to chemical attack, though it can be washed with care. Uncoated mirrors can also be washed, but this does little to restore reflectivity in the UV. More recently silica has been used as a protective coating and although this does not prevent contamination, cleaning the surface can restore 'as-new' performance. This is shown in Fig. 3.7 demonstrating that the silica coating prevents the formation of an oxide layer, and that the accumulated surface debris can be completely washed off. The cleaning of optical components is discussed further in Chapter 10. Silica-coated mirrors also exhibit a remarkable 'self-cleaning' effect after about 200 days irradiation, which can restore their performance almost completely for up to a further 150 days [14].

Another more flexible – in every sense of the word – approach to the problem of getting light round the optical path of the instrument is to use fibre optic light guides. Although the UV transmission

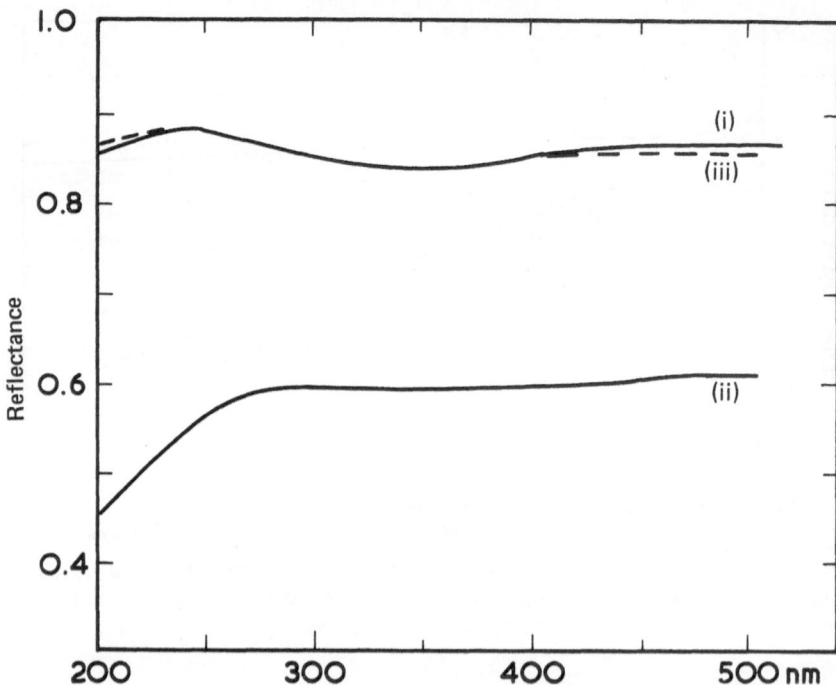

Fig. 3.7 *The reflectivity of silica-coated aluminium mirror: (i) before irradiation; (ii) after 530 days UV irradiation; (iii) after washing. This shows that the loss of reflectivity is almost entirely due to surface contamination. Drawn from data in [13].*

characteristics of silica fibre optics are not particularly impressive, they can be used with a high intensity pulsed light source to make a viable system. The approach with this type of instrument is to take the light to the sample rather than the other way round, and there is at least one manufacturer who produces a spectrometer fibre optic light probe which can be used to measure either absorption, reflection, fluorescence or phosphorescence of sample solutions remote from the instrument in much the same way as a pH electrode is used.

References

1 Allan, W.B. (1973), *Fibre Optics Theory and Practice*, Plenum Press, London and New York.
2 Hughes, D. (1977), *Chemistry in Britain*, **13**, 134.
3 Health and Safety Executive (1978), Guidance Note EH 15/78, Threshold Limit Values for 1978, H.M. Stationery Office.

4 Tolles, W.M., Nibler, N.W., McDonald, J.R. and Harvey, A.B. (1977), *Appl. Spectrosc.*, **31**, 253.
5 Turk, G.C., Travis, J.C., DeVoe, J.R. and O'Haver, T.C. (1978), *Anal. Chem.*, **50**, 817.
6 Wright, J.C. and Wirth, M.J. (1980), *Anal. Chem.*, **52**, 1087A.
7 Wright, J.C. and Wirth, M.J. (1980), *Anal. Chem.*, **52**, 989A.
8 Maiman, T. (1960), *Nature*, **187**, 493.
9 West, M.A. (1979), *Developments in Instrumentation and Techniques in Photochemistry*, Vol. 10, Specialist Periodical Report, Chemical Society, London.
10 Bilt, S., Neaver, E.G., Rabson, T.A. and Tittle, F.K. (1978), *Appl. Optics*, **17**, 721.
11 Shirk, J.S., Harris, T.D. and Mitchell, J.W. (1980), *Anal. Chem.*, **52**, 1701.
12 Dovichi, N.J. and Harris, J.M. (1979), *Anal. Chem.*, **51**, 728.
13 Francis, R.J. (1980), *International Laboratory*, May/June, 85.
14 Francis, R.J. (1978), *UV Spectrom. Grp Bull.*, **6**, 35.

General reading

Rabek, J.F. (1982), *Experimental Methods in Photochemistry and Photophysics*, John Wiley & Sons, Chichester, pp. 34–212.

4 Monochromators

The monochromator is in essence the heart of any spectrometer. On it depend such fundamental parameters as wavelength accuracy and resolution. In general, it contains a system of slits and lenses or mirrors, together with a dispersing element which may be either a prism or grating. Before we can consider the detail design of complete monochromators we need to understand the fundamentals of the basic components.

4.1   Prisms

For many years, prisms formed the basis of all good spectrometers. A schematic diagram of a simple prism monochromator is shown in Fig. 4.1. Refraction occurs at both faces of the prism, and the dispersed radiation is then focused on a slightly curved surface containing the exit slit. Radiation of the desired wavelength is then obtained by rotation of the grating. The spectral purity of the radiation emerging is determined mainly by the dispersion characteristics of the prism. In Fig. 4.2, the angular dispersion is $d\theta/d\lambda$ which may be expressed as

$$\frac{d\theta}{dn}\frac{dn}{d\lambda}$$

where θ is the angle of deviation of a ray of wavelength λ, and n is

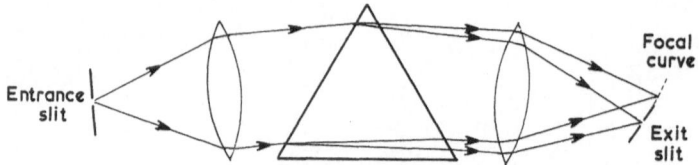

Fig. 4.1 *A prism monochromator. Redrawn from Skoog and West [20].*

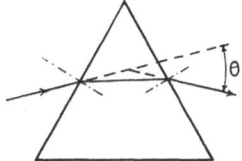

Fig. 4.2 *Dispersion by a prism. Redrawn from Skoog and West [20].*

the refractive index. The term $d\theta/dn$ is a function of the geometry of the prism and the angle of incidence. A full description can be found in James and Sternberg [1] and Stewart [2]. The term $dn/d\lambda$ is related to the dispersion of the material from which the prism is constructed. Typical curves for a number of materials commonly used for prisms are shown in Fig. 4.3. This figure contains the key to the demise of prisms and dispersing elements. The complex relationship between refractive index and wavelength results in a very non-linear distribution of wavelengths along the plane of the exit slit. For this reason the wavelength control of a prism monochromator is usually indirectly coupled to the prism through an accurately machined cam and lever mechanism to give a linear wavelength scale.

Another problem of a prism is the fact that the angular dispersion $d\theta/d\lambda$ is a function of temperature. Therefore, in order to avoid wavelength setting inaccuracies due to changes in ambient temperature, a good prism monochromator uses either a bimetallic link in

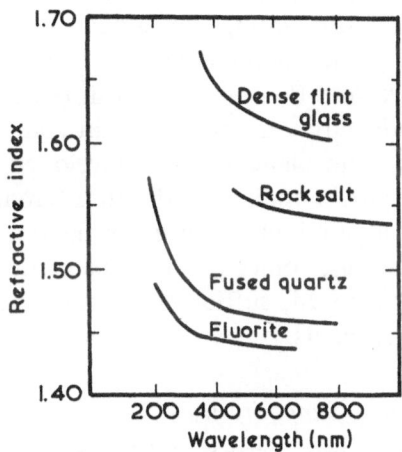

Fig. 4.3 *The dispersion of various optical materials. Redrawn from Skoog and West [20].*

the wavelength selection mechanism or the whole monochromator is thermostatted at a temperature above ambient. If a wavelength scan is to be made with a prism instrument, and results are required to be made at constant bandwidth, then the width of the slit must be adjustable so that it is narrow at wavelengths where the spectrum from the prism is bunched up, and wide where the spectrum is spread out.

It will be appreciated from these considerations that although a basic prism is fairly simple to manufacture, the mechanical problems of incorporating it into a workable monochromator can be quite formidable. This is one of the reasons why grating monochromators are now preferred by most manufacturers.

4.2   Gratings

A diffraction grating is a very simple optical device, but until recently, has been very difficult to produce. The traditional method has been by means of a ruling engine. For typical UV–VIS spectrometers, 1200 lines/mm are required over an area of 20 mm x 20 mm. Hutley [3] expresses the problem as being equivalent to ploughing a furrow 10 km long with an overall straightness of 20 nm and a short term deviation of 1 nm. There are few centres in the world with this capability, and so replica gratings are made from the ruled masters. These are produced in rather the same way that gramophone records are pressed from a master, but in a more refined manner.

More recently, holographic diffraction gratings or more properly, interference diffraction gratings, have been introduced into spectrometers. These are manufactured by spinning a layer of photoresist onto a glass blank, and then exposing this photoresist to an interference pattern produced by a coherent light source which is usually a high power laser. The blank is then 'developed' to leave accurately parallel grooves corresponding to the interference pattern, which is overcoated with aluminium to form a reflection diffraction grating. As with ruled gratings, copies can then be made if required, or if the process is well organized, sufficient masters can be produced to fit originals into instruments.

4.2.1   *Diffraction grating physics*

A simple understanding of the physics of diffraction gratings will help us understand some of the principles of monochromator design.

The basic diffraction grating equation

$$k\lambda = d\ (\sin\theta + \sin\beta) \tag{4.1}$$

can be easily derived (Fig. 4.4) and is explained in many text books by saying that constructive interference between the reflected light from adjacent facets will occur when the path difference $= k\lambda$, where k is the diffracted order (usually a small integer). This point will be covered again when we consider second-order spectra in monochromators. An important case of Equation (4.1) is when $\theta = \beta$, i.e. the Littrow mounting, so the equation reduces to

$$k\lambda = 2d\sin\theta \tag{4.2}$$

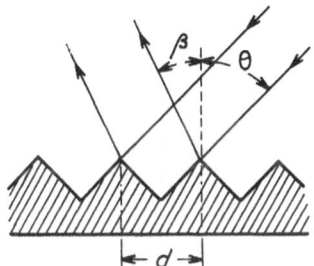

Fig. 4.4 *Diffraction at a reflection grating. Redrawn from Skoog and West [20].*

4.2.2   Dispersion

Angular dispersion is the angular separation $d\beta$ obtained for two different wavelengths separated by $\Delta\lambda$, and can be obtained by differentiating Equation (4.1):

$$\cos\beta\ d\beta = \frac{k}{d}\ d\lambda$$

$$\frac{d\beta}{d\lambda} = \frac{k}{d\cos\beta} \tag{4.3}$$

Linear dispersion is the product of this and the effective focal length of the monochromator. Usually in a UV–VIS instrument one wishes to keep the focal length as short as possible to produce a compact instrument, and so to obtain as much linear dispersion as possible, it is tempting to aim for high angular dispersion. Equation (4.3) should, however, be treated with great care, as the ratio k/d is not, in fact, independent of wavelength.

If Equation (4.1) is solved for k/d and the result substituted in Equation (4.3) the general equation for angular dispersion is obtained

$$\frac{d\beta}{d\lambda} = \frac{1}{\lambda} \frac{\sin \theta \sin \beta}{\cos \beta} \tag{4.4}$$

i.e. angular dispersion is a function of the angles of incidence and diffraction. A fuller description of this aspect of optical design can be found in [4] and [5].

4.2.3 *Resolution*

The spectral resolution of a monochromator is its ability to separate adjacent spectral lines, and is generally defined as $R = \lambda/d\lambda$. The standard textbook derivation of this for a diffraction grating leads to the expression

$$R = kN$$

where k is the order and N the total number of lines on the grating. This equation must be treated with care. If the width of the grating is fixed, then one is tempted to increase the number of lines per nanometre in order to increase N, but this can itself lead to other problems. In practice, in UV—VIS spectroscopy, grating resolution is seldom a limiting factor in the design of monochromators.

4.2.4 *Blaze*

The distribution of light energy amongst the various orders (given by various values of k in Equation (4.1)) depends strongly on the shape of the individual grooves: for a sinusoidal grating, the energy distribution is as shown in Fig. 4.5a. In most monochromator designs the grating is used in first-order, and so an awful lot of energy would be wasted in the other orders if a sinusoidal grating were used.

The trick is to change the angle of the reflecting surfaces relative to the plane of the grating surface (Fig. 4.5b). This technique is known as blazing and has the effect of moving the envelope of Fig. 4.5a sideways. The blaze wavelength is the wavelength at which the grating has maximum efficiency. A fuller description of the mathematics of blaze and efficiency can be found in a number of references [4–7]. Efficiency curves for a number of typical diffraction gratings are given in Fig. 4.6. This diagram illustrates a number of interesting points. Notice the peculiar irregularities around 600 nm for the blazed ruled grating. These are known as Wood's anomalies and are

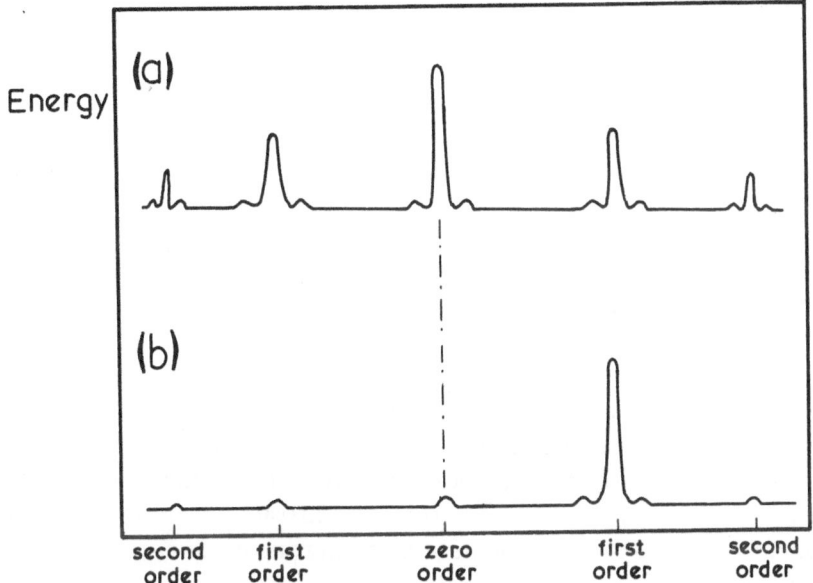

Fig. 4.5 *(a) The energy profile resulting from the reflection of a beam of mono-chromatic light from a grating. Most of the energy is concentrated in the non-dispersed zero-order band, while decreasing amounts appear in the pairs of first- and higher-order bands; (b) the improved energy profile from a blazed grating where most of the energy is concentrated in a single first-order band.*

polarization-dependent. They can be the cause of some peculiar effects, such as baseline irregularities, in instruments. The two curves for the holographic gratings show the difference in efficiency that can be achieved by blazing the grating. Note the reduced size of the Wood's anomalies.

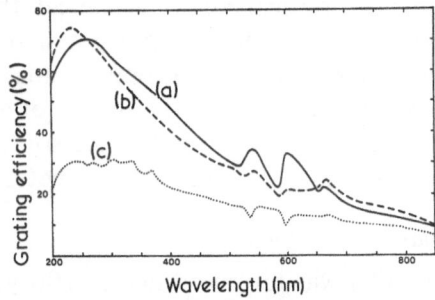

Fig. 4.6 *The variation in the efficiency of different types of grating with wave-length: (a) ruled grating blazed at 240 nm; (b) holographic grating blazed at 260 nm; (c) sinusoidal holographic grating. Reproduced with permission from Pye Unicam Ltd.*

It can also be seen that the gratings have been blazed for around 250 nm. This is important in the design of UV–VIS instruments where the signal coming from the detector is the convolution of lamp energy x mirror reflectivities x grating efficiency x detector sensitivity. If the blaze were moved into the visible, the signal-to-noise ratio at 200 nm would suffer, and if it were moved more to the UV, signal-to-noise at the longwave end of the spectrum would suffer.

4.2.5   *Stray-light*

This topic has been covered extensively in Volume 1 of this series [8]. It is worth noting here that in a well-designed optical system, the grating has traditionally been the major source of stray-light. It has been stated [9] that three basic errors exist in gratings: groove depth variations, groove spacing errors, and groove straightness can give rise to stray-light. With the introduction of holographic gratings the contribution to scattered light from the diffraction grating is much reduced, with other components (e.g. the grating mount) contributing significantly to the total stray-light.

Fig. 4.7 shows electron micrographs of the three types of grating discussed, and partly illustrates why blazed holographic gratings are superior to conventional ruled gratings. This is further illustrated in Fig. 4.8, which shows the stray-light levels of the various gratings when illuminated with light from a helium–neon laser. The light level between the zero-order and the first-order is a measure of their relative stray-light performance in a spectrometer.

4.3   Monochromator design

The design of monochromators is a specialist subject which could fill many pages with optical and mechanical theory. As well as designing a compact system, it is important that the main primary optical aberrations (spherical aberration, coma, astigmatism) are minimized. A general treatment on the theory of aberrations is given by Jenkins and White [10]. Given below are some of the more common arrangements found in UV–VIS spectrometers.

4.3.1   *The in-plane Ebert mounting*

This system (Fig. 4.9) was proposed nearly 100 years ago by Ebert, and was popularized by Fastie in 1952 [11] when he described a monochromator based on this system and enumerated some of its advantages. The elegance of the system is two-fold. First, although

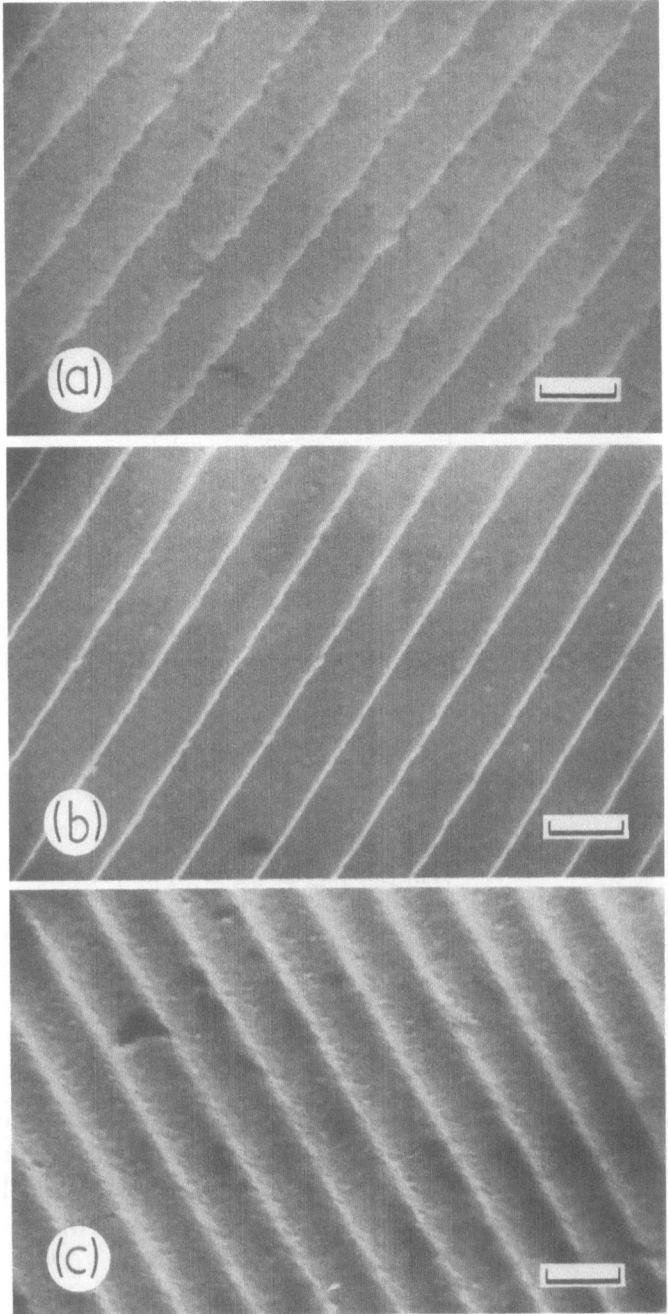

Fig. 4.7 *Scanning electron micrographs of the surface of diffraction gratings of different types: (a) a blazed ruled grating; (b) a blazed holographic grating — the improved regularity and smoothness is apparent; (c) a sinusoidal holographic grating. The bars represent 1 μm. Reproduced with permission of Pye Unicam Ltd.*

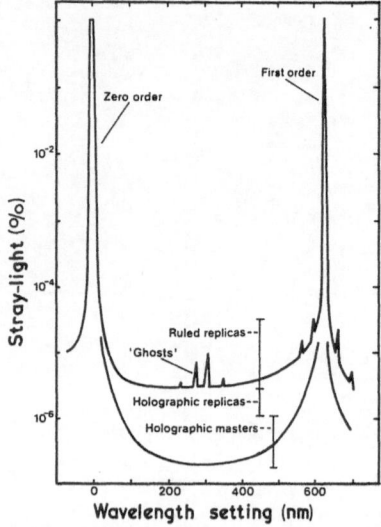

Fig. 4.8 *Stray-light profiles for different types of grating. The bars represent the range of values obtained with the different gratings at the minimum between the orders. Stray-light was measured at 633 nm. Reproduced with permission from Pye Unicam Ltd.*

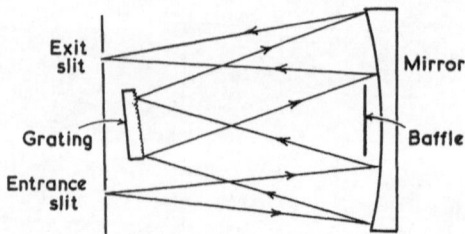

Fig. 4.9 *The in-plane Ebert mounting.*

the mirrors are used off-axis, the primary coma introduced by the first mirror is mostly removed by the opposite coma of the second mirror, although it should be noted that as the Ebert angle becomes larger, the correction holds less well. Secondly, the use of symmetrical curved slits minimizes the effects of astigmatism. Variants of this basic in-plane Ebert system include the earlier Czerny–Turner system where two separate mirrors are used instead of the single mirror, and the off-plane Ebert in which the slits are arranged above and below the grating. These are described at length by James and Sternberg [1].

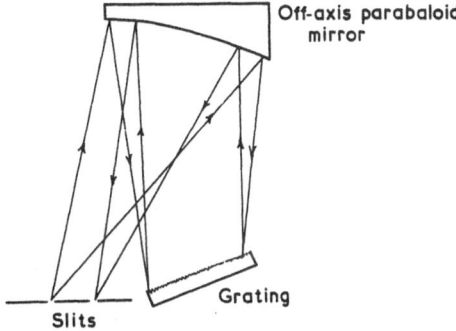

Fig. 4.10 *The Littrow mounting.*

4.3.2 *The Littrow mounting*

The beauty of the Littrow mounting is really its simplicity. A single mirror, either spherical, or preferably an off-axis paraboloid, is used as shown in Fig. 4.10. This leads to a very compact and cheap system for it can be seen that the whole monochromator need be little wider than the width of the grating. A detailed analysis of the aberrations in the Littrow system, and of its limitations is given by Kudo [12].

4.3.3 *The Monk–Gillieson mounting*

This rather unusual design employs a plane grating in a converging beam of light as shown in Fig. 4.11. It has been well documented in a number of papers [13–15]. The aberrations are such that it is not really suited for use in narrow bandpass instruments. The optical layout, however, is attractive as the lamp house can be at one side of the monochromator and the sample compartment at the other. In the designs described previously, a corner mirror has always been required to direct one of the light beams out of the way. It tends to find application, therefore, in single-beam instruments, with a bandpass of typically 8 nm.

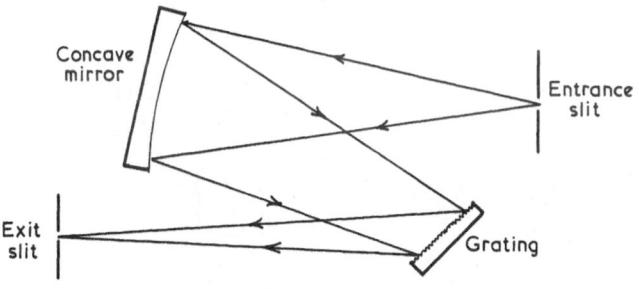

Fig. 4.11 *The Monk–Gillieson mounting.*

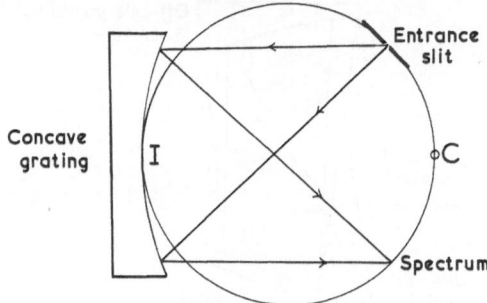

Fig. 4.12 *A concave grating and the Rowland Circle.*

4.3.4   *Concave gratings and their mountings*

In theory concave gratings are very attractive for use in UV–VIS spectrometers. They combine in one optical element the collimating optics, the grating, and the focusing optics. Their chief disadvantage is the inherent astigmatism [10] which limits their use at narrow bandwidth.

The key to understanding concave gratings lies in the Rowland Circle (Fig. 4.12). In the diagram, C is the centre of curvature of the grating and I is the vertex of the grating. Rowland showed in 1882 that if the entrance slit lies on the circle whose diameter is IC, then the spectrum also lies on this circle. A typical concave grating mono-chromator is shown in Fig. 4.13. This is the Seya–Namioka design.

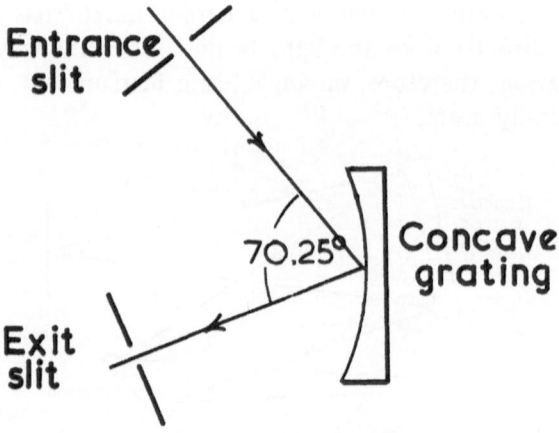

Fig. 4.13 *The Seya mounting.*

There are many more designs and solutions, mostly originating from Japanese designers (see, for example, [16–19]).

The disadvantage with all the designs continues to be astigmatism, although as shown by Stroke [6] it is in fact possible to design a concave holographic grating which is completely stigmatic at three given wavelengths. In addition, the angle between entrance and exit slit makes the design of the rest of the optical system rather awkward, without the addition of further mirrors. However, if more mirrors are introduced, much of the benefit of the concave system, namely fewer optical components to produce and align, has been lost.

4.4 Double monochromators

A number of instruments now incorporate a pre-monochromator (sometimes referred to as a 'fore-monochromator'). This second monochromator is placed immediately before the main monochromator so that the exit slit of the pre-monochromator also serves as the entrance slit of the second monochromator. The role of the pre-monochromator is basically to reduce the stray-light of the instrument. This improvement is, however, bought at a price. Fig. 4.6 gave

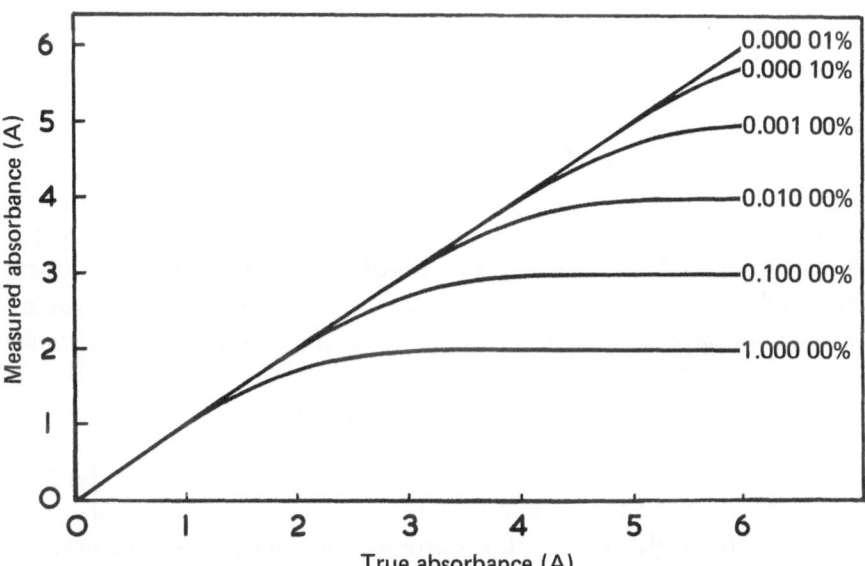

Fig. 4.14 *The effect of stray-light on the apparent photometric linearity of a spectrometer. The numbers are the percentages of stray-light corresponding to each curve. Reproduced with permission from Pye Unicam Ltd.*

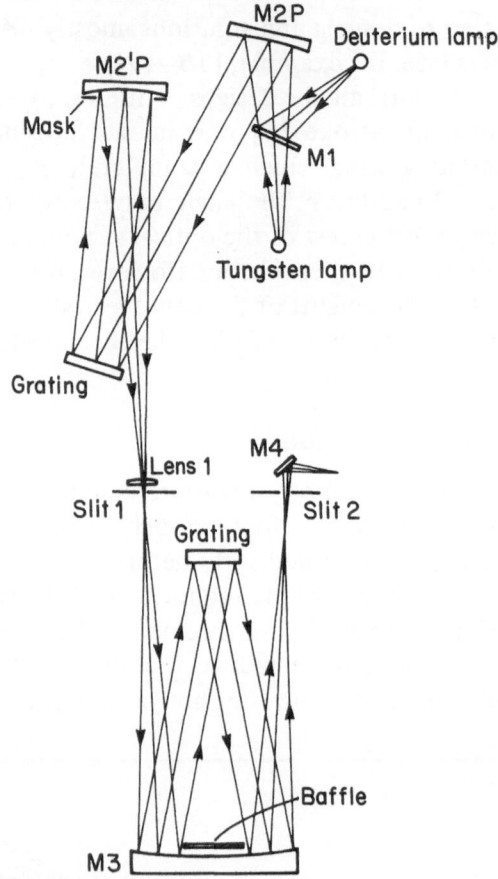

Fig. 4.15 *The addition of a pre-monochromator to an Ebert monochromator.*

a typical efficiency curve for a grating. If a second grating with the same efficiency is introduced into the system, it can be seen that the light throughput, especially at the ends of the spectrum, is dramatically reduced, with a corresponding increase in signal-to-noise. A double-monochromator instrument should therefore be the choice of those users consistently working at high absorbance values where stray-light may be important. Fig. 4.14 illustrates the effects of stray-light on linearity at high absorbances.

In general, the design criteria for pre-monochromators are far less stringent than for the main monochromator. Thus the lamp filament can be effectively used as the entrance slit, and a concave grating may be used as the dispersing element. The motion of the two gratings is usually ganged together by a simple mechanical linkage. A typical layout is shown in Fig. 4.15.

4.5   Second-order and stray-light filters

Referring back to Equation (4.1), it will be seen that for different integral values of k, different values of λ will satisfy the equation. Thus when the instrument is set to transmit light of wavelength λ, there will be a certain amount of radiation of wavelength $\lambda/2$ transmitted in second order, $\lambda/3$ in third-order and so on. These wavelengths must be excluded by some form of filtering, which is usually achieved by the use of a suitable coloured glass filter which is automatically inserted into the beam at the appropriate wavelength. In most UV–VIS instruments a red transmitting (blue absorbing) filter is inserted at wavelengths of ~ 680 nm and above, thus stopping second-order light of 340 nm and above being transmitted at the same time. Below this wavelength, the output from the tungsten lamp is dropping rapidly, and so higher orders do not need to be filtered out.

A number of stray-light filters may also be inserted in the beam. The reason for these can be partly explained by Fig. 4.16. This shows a typical 'energy profile' of a spectrometer, i.e. the product of lamp output x (mirror reflectivity)^n x grating efficiency x detector response, where n is the number of mirrors in the system. Thus if the monochromator is set to 340 nm, the effect of any stray-light at, say, 500 nm will be greatly magnified by the ratio of the energy profile at the two wavelengths. A filter which transmits at 340 nm, but blocks light of higher wavelengths will therefore greatly improve the measured stray-light of the instrument.

The number of filters used, and the wavelengths at which they are inserted varies from instrument to instrument and is really dependent

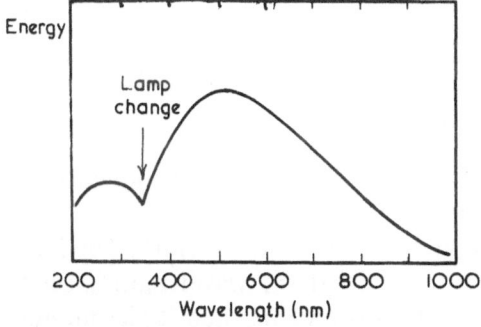

Fig. 4.16 *The energy profile of a typical spectrometer with deuterium and tungsten lamps. This represents the product of lamp output, mirror reflectivities, grating efficiency and detector spectral response.*

on how well the monochromator has been designed and on the quality
of the diffraction grating used.

4.6 Slits and slit mechanisms

The slits play an important role in determining the quality of the
monochromator. It is important that the edges are exactly parallel to
one another and that they lie in the same plane. As the width of the
slit may be only 20 μm or so, it must be made with great precision.
It is also good design practice to reduce to a minimum the thickness
of material at the edges, so the least amount of light is scattered
inside the monochromator (Fig. 4.17).

In older instruments the width of the slits had to be continuously
variable and a good description of the many types of slit mechanisms
can be found in [1]. Modern instruments tend to offer the user a
series of discrete slitwidths. These individual slits form part of a disc
which can be actuated by a stepper motor. The slits are usually
formed by a chemical etching technique.

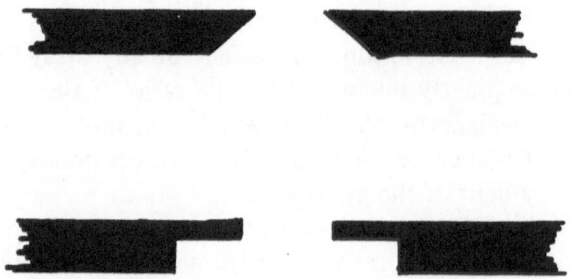

Fig. 4.17 *Typical slit jaw profiles.*

4.7 Bandwidth

Fig. 4.18 illustrates the definition of bandwidth or, better, effective
spectral slitwidth (ESW). A monochromator is set up so that the
entrance and exit slits are of equal width. The image of the entrance
slit just fills the exit slit. If the wavelength is set to λ_2 and the slitwidth is equivalent to the ESW, then if the monochromator is reset
to λ_1 or λ_3, the image of the entrance slit is moved completely out
of the exit slit. The energy distribution as the image of the entrance

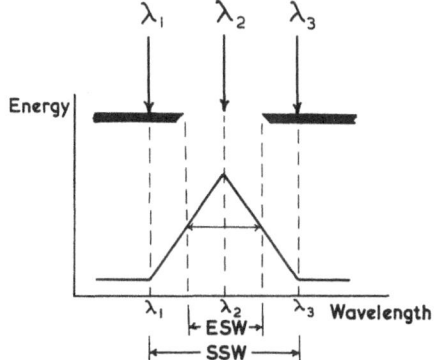

Fig. 4.18 *Spectral profile of the energy leaving a monochromator exit slit. The monochromator is set to λ_2 but, in fact, wavelengths ranging from λ_1 to λ_3 leave the exit slit. The distribution is roughly triangular if the slit is small. The total spread of wavelength, $\lambda_3 - \lambda_1$, is termed the spectral slitwidth (SSW), and the width of the profile at half-peak height is termed the effective spectral slit-width (ESW).*

slit is moved across the exit slit is shown below, and simply shows that the convolution of two rectangular functions is a triangular function. The spectral bandwidth is also shown as the bandwidth at half-peak intensity.

4.8   Monochromator drive mechanisms

Most prism monochromators and some grating monochromators are driven by a cam arrangement (Fig. 4.19a). The cam is so machined that a linear wavelength scale results. Accurate cams are, of course, expensive to produce, and most grating instruments are now driven by a sine bar mechanism (Fig. 4.19b). If the instrument covers a range of 200–900 nm and wavelength is to be set to within ± 0.2 nm, a repeatability of 1 in 3500 is called for, over a distance which is typically 20 mm. Thus a micrometer screw arrangement driven by a suitably geared stepper motor is used.

4.9   Filter instruments

For those analyses where spectral purity of the radiation is not important, or where an inexpensive instrument is a necessity, then often a simple colorimeter will suffice. The very simplest form is a visual colorimeter, using the human eye as the detector. More com-

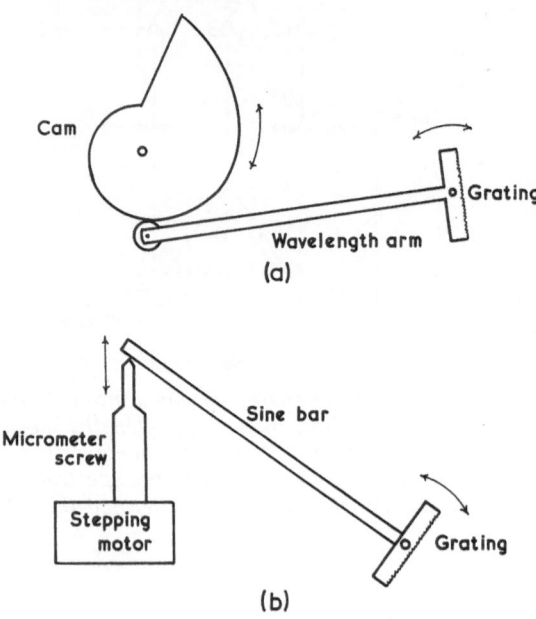

Fig. 4.19 *Grating drive mechanisms: (a) driven by a specially designed cam; (b) the 'sine bar' arrangement.*

monly, a photoelectric colorimeter is used. This consists of a tungsten lamp, focusing element (lens or mirror), filter, sample holder and detector for which a simple photoelectric cell will suffice. To use the instrument, a cell containing an appropriate blank is placed in the beam, and the instrument readout zeroed. This may be accomplished either by a mechanical attenuator or by varying the voltage to the lamp. The sample is then placed in the beam and the appropriate reading noted. Instruments are generally supplied with a number of filters and the selection of the appropriate filter is very important as the sensitivity of the measurement is dependent upon the filter – typically it will have a bandpass of 40 nm compared to a bandpass of between 1 and 10 nm in a monochromator instrument. Remember that a solution appears green because it transmits the green portion of the spectrum, but absorbs the red. It is the intensity of the red radiation which varies with concentration, and a red filter should therefore be used. As a general rule the filter should be the colour complement of the solution being analysed. Colorimeters, therefore, are most widely used for samples which do not exhibit very sharp absorption bands and in those applications where absolute accuracy is not essential.

References

1　James, J.F. and Sternberg, R.S. (1969), *The Design of Optical Spectrometers*, Chapman and Hall, London.
2　Stewart, J.E. (1970), *Infrared Spectroscopy*, Marcel Dekker, New York.
3　Hutley, M.C. (1974), *Science Progress (Oxford)*, **61**, 301.
4　Anon. (1970), *Diffraction Grating Handbook*, Bausch and Lomb, Rochester.
5　Anon. (1972), *Diffraction Gratings; Ruled and Holographic*, Jobin Yvon Inc., Longjumeau.
6　Stroke, G.W. (1967), *Diffraction Gratings*, in *Encyclopedia of Physics*, Vol. XXIX *Optical instruments* (Ed. S. Flügge), Springer-Verlag, Berlin.
7　Hutley, M.C. (1982), *Diffraction Gratings*, Academic Press, London.
8　Burgess, C. and Knowles, A. (1981), *Techniques in Visible and Ultraviolet Spectrometry*, Vol. 1, *Standards in Absorption Spectrometry*, Chapman and Hall, London.
9　Palmer, E.W., Hutley, M.C., Franks, A., Verrill, J.F. and Gale, B. (1975), *Rep. Prog. Phys.*, **38**, 975.
10　Jenkins, F.A. and White, H.E. (1957), *Fundamentals of Optics*, McGraw Hill, New York.
11　Fastie, W.G. (1952), *J. Opt. Soc. Amer.*, **42**, 641.
12　Kudo, K. (1960), *Science of Light*, **9**, 65.
13　Monk, G.S. (1928), *J. Opt. Soc. Amer.*, **17**, 358.
14　Gillieson, A.H.C.P. (1949), *J. Sci. Inst.*, **26**, 335.
15　Schroeder, D.J. (1966), *Appl. Optics*, **5**, 545.
16　Seya, M. (1952), *Science of Light*, **2**, 8.
17　Namioka, T. (1959), *J. Opt. Soc. Amer.*, **49**, 460.
18　Onaka, R. (1958), *Science of Light*, **7**, 23.
19　Welford, W.T. (1964), *Aberration Theory of Gratings and Grating Mounts* in *Progress in Optics*, Vol. IV, North Holland, Amsterdam.
20　Skoog, D.A. and West, D.M. (1971), *Principles of Instrumental Analysis*, Holt, Rinehart and Winston, New York.

5 Detectors

5.1   Introduction

As a user of a routine spectrometer you will probably take the detector for granted, giving it very little thought. This is largely as it should be since the instrument manufacturer will have selected a detector consistent with the price and performance target of the spectrometer, and the user is not involved in setting detector conditions as he is in selecting, say, slitwidth and scanning speed. In simple, manual spectrometers using low cost, restricted wavelength range detectors, it may be necessary to select the appropriately sensitive unit, but this is about the extent of the user's involvement. Moreover, the high reliability of modern detectors means that they require no routine attention and failures are rare.

Despite the foregoing, it is valuable to know something of the design and operation of the detectors used in spectrometers in order to understand the limits of performance you can expect from the technique which are determined by this component. Although the purpose of the detector may seem self-evident, i.e. to turn a level of illumination into an electrical signal so that the sample's transmission or reflectance can be measured, the type of detector used will be determined by the overall purpose of the spectrometer.

There are several mechanisms employed to generate electrical signals from light in spectrometer detectors. Most instruments use vacuum tube devices whilst various solid state types are used in the remainder, and may become predominant in future. The types discussed in this chapter will be: (a) vacuum tube detectors — phototubes and photomultipliers; (b) solid state devices — barrier layer photocells, lead sulphide cells, silicon diodes, diode arrays and Vidicon tubes.

5.2 Vacuum tube detectors

5.2.1 *Phototubes*

The phototube is used in the vast majority of manual single-beam spectrometers. It offers good sensitivity enabling modest slitwidths to be employed, has long life, good stability and requires simple electronics.

The phototube consists of a photocathode, usually in the form of a half cylinder with an anode rod at its axis. The whole is enclosed in a glass envelope, or silica if shortwave sensitivity is desired, and looks like a radio valve. The usual dimensions are about 50 mm long by 20 mm diameter although sizes vary considerably. The cathode is covered with a thin layer of alkali metal, usually caesium alloyed with antimony and often including silver and silver oxide. By choosing from various coatings, peak spectral sensitivity is varied from the UV to VIS regions. Fig. 5.1 shows the sensitivity of various photocathode metals.

When light strikes the cathode, photoelectrons are emitted and will be drawn to the anode, usually held about 90 V positive with respect to the cathode. The more intense the light, the greater the photocurrent. This photocurrent, as small as 10 pA, is then amplified but remains effectively proportional to the incident light level up to high illumination levels, and thus the signal is proportional to sample

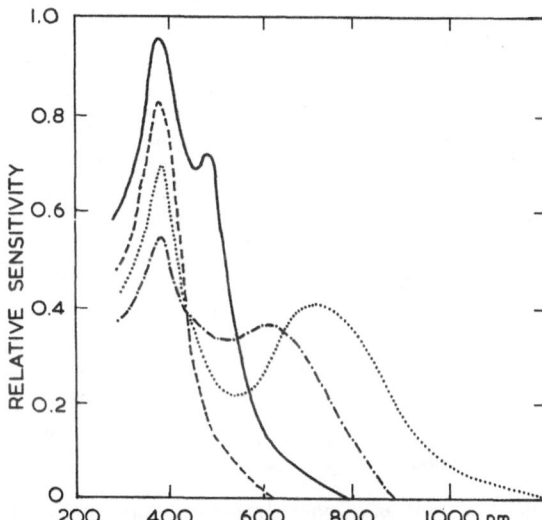

Fig. 5.1 *Spectral sensitivities of different alkali metal photocathodes: potassium (——); sodium (– – –); rubidium (– · – ·); caesium (. . .).*

transmittance once the 'unit transmittance' control has been set. Absorbance readout is derived by a logarithmic converter either electronically or via a microprocessor. Since there is also a random emission of thermal electrons the phototube will produce a current even when unilluminated called the 'dark' current. Although small it must be compensated either by initially backing off the amplifier with no light entering the detector, or providing a dark sector in the photometer chopper for continuous monitoring.

5.2.2 Photomultipliers

The photomultiplier, used in most double-beam scanning spectro-meters, exceeds the sensitivity of the simple phototube by about 200-fold whilst maintaining its relative ease of use, stability and long life. This is achieved by adding photoelectron multiplication stages to the phototube. Fig. 5.2a and b illustrate the construction of side-window and end-window photomultipliers. A photocathode similar to that of the phototube emits photoelectrons when illuminated but, in this case, instead of collecting the current at an anode, a further

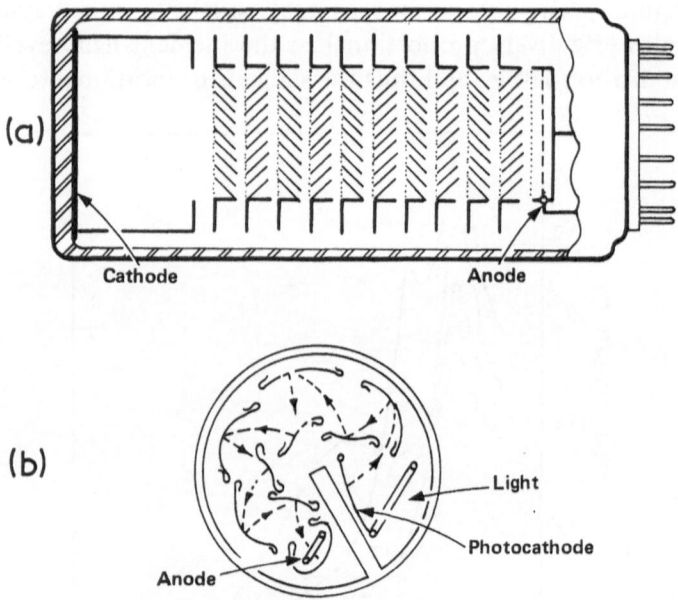

Fig. 5.2 *Two types of photomultiplier: (a) longitudinal section of an end-window photomultiplier with dynodes in the 'venetian blind' conformation; (b) transverse section of a side-window photomultiplier with dynodes arranged for circular focusing. Reproduced by permission from Thorn–EMI Ltd.*

electrode called a dynode, which is held at about 70—100 V positive with respect to the cathode, attracts the photoelectrons. The surface of the dynode is coated with antimony—caesium or beryllium—copper alloys and emits several secondary electrons, usually four or five for each photoelectron that strikes it, and thus amplifies the photo-electric current proportionally. These emitted electrons are attracted to a second dynode with a further current increase and so on for from six to ten dynodes. After final collection at the anode an overall gain of between 10^6 and 10^7 has been achieved.

Although not as sensitive as the end-window photomultiplier, the side-window type is most frequently used in scanning spectrometers due to its compactness (Fig. 5.3). The end-window variety finds its application where ultra-high sensitivity is required or where the large end window optimizes the collecting of diffuse radiation from turbid samples. The spectral response and gain of the photomultiplier is controlled by the photocathode layer, dynode layer and window material, and their overall responses are characterized by manufactur-ers' 'S' numbers as shown in Fig. 5.4 and Table 5.1. These 'S' numbers qualitatively describe the response of a photocathode and are similar to the 'EIA' numbering system of the American Electrical Industries Association. The 'R' numbers used by Hamamatsu are also similar.

To achieve their high gain, photomultipliers require a dynode supply voltage in the region of 1 kV, which should be regarded as the price of high sensitivity. This voltage must be very stable to ensure drift-free operation of the detector. Furthermore, the dark current limits the maximum practical working voltage: sensitivity can be increased by raising the dynode voltage, but you cannot indefinitely increase it to amplify low-level light signals because the dark current is also amplified and will eventually dominate the output. The response time of the photomultiplier is very short permitting pulse rates of 10^9 s^{-1}, so high speed chopping can be used when following absorb-ance changes although normal chopping frequencies are several orders of magnitude slower than this.

Like the phototube, response is linear up to high light levels and so a signal proportional to sample transmittance can readily be produced. However, beware! Owing to the very high gain, excessive light reach-ing the detector, i.e. daylight or even bright laboratory illumination, will have dramatic or even disastrous effects on a working photo-multiplier. The super high abundance of photoelectrons produced under such circumstances can either saturate the detector, requiring several hours to recover normal performance, or cause permanent

Fig. 5.3 *A side-window photomultiplier. Reproduced by permission from Thorn–EMI Ltd.*

damage to the photocathode and dynode surfaces. To avoid this problem most spectrometers either have a solenoid or mechanically operated shutter which interrupts the light beam, or a light sensor which will cut down the dynode voltage should the sample compartment be opened or the instrument cover be removed.

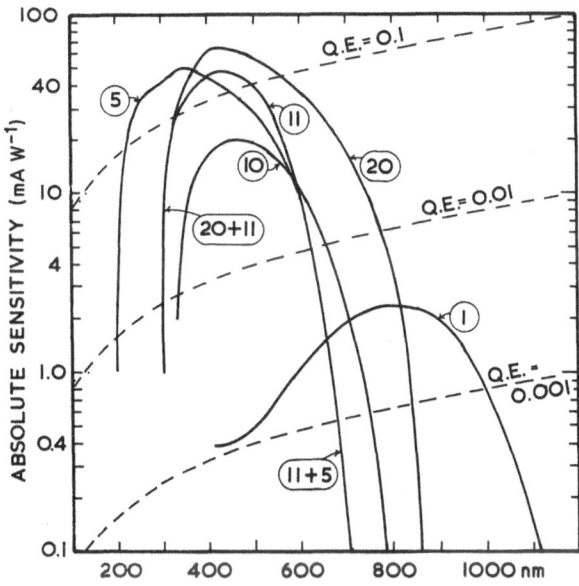

Fig. 5.4 *Spectral sensitivity curves for various types of photocathode. Note that the vertical scale is logarithmic. The lines marked Q.E. are the curves for cathodes of quantum efficiency 0.1, 0.01 and 0.001 electrons per photon. Redrawn from data supplied by RCA Ltd.*

Table 5.1: *A selection of photo-cathode types.*

Cathode type	Composition	Peak sensitivity (nm)
S1	AgOCs	800
S5	SbCs	300
S10	BiAgOCs	420
S11	$Cs_3 SbO$	390
S20	$Na_2 KSBCs$	300

5.3   Solid-state detectors

5.3.1   *Barrier layer photocells*

This is a very simple photovoltaic device which generates a voltage when its semiconductor-type surface is illuminated. It consists of a layer of selenium a few microns thick sandwiched between a thick base electrode of iron or aluminium and a sputtered surface coating of silver or gold which faces the illumination. Radiation passes through the very thin sputtered silver coating and promotes electrons at the silver—selenium interface to a conducting state; these electrons then migrate to the collector electrode on the silver surface. Thus a voltage is generated between the base and collector electrodes which is sufficient to drive a galvanometer or microammeter with a resistance of 400 ohm or less. The current is proportional to light intensity so a transmittance reading is readily obtainable.

The sensitivity of this device is quite low and restricted to the visible region. With a slow response and a tendency to fatigue, it is suitable only for the most routine spectrometers and its use declined as a result.

5.3.2   *Lead sulphide cells*

These are semiconductor devices which change their conductance on illumination. A number of semiconductor materials based on the selenides and stibnides of lead, cadmium, gallium, and indium have been used but lead sulphide is the most frequently encountered with good sensitivity in the 750 nm to 3 μm region. This gives a usefully sensitive device for the near-IR region as it covers the gap between the red-sensitive photomultipliers and the thermocouples used in IR spectrometers. The crystalline semiconductor is coated on glass and sealed in an evacuated container to prevent atmospheric deterioration. Illumination causes some boundary electrons to be promoted to a conducting state and this conductivity increase can be measured in a bridge circuit.

5.3.3   *Silicon diode detectors*

The silicon diode is formed on a silicon slice or chip and consists of a reverse-biased pn-junction. The reverse bias produces a practically zero conductance which increases on illumination as the holes from the electron—hole pairs formed in the n-layer migrate to the p-layer and are annihilated. Thus the conductance is proportional to the level of illumination. The silicon diode is less sensitive than the photomultiplier, but of course, does not require the high voltage supply. Moreover, it is very small and individual units are packaged like a transistor, making it ideal for compact optical systems. Good sensitivity is shown in the UV and VIS regions.

5.3.4   *Diode arrays*

The small physical size of the silicon diode opens up some interesting possibilities, one of which is now available in a commercial instrument.

By placing an array of diodes in the dispersed radiation from a grating it is possible to monitor all wavelengths simultaneously in a spectrometer which uses white light to illuminate the sample. This system obviates the need to move the grating, spectral scans are obtained by scanning the photodiodes electronically, and this rapid parallel detection permits high signal-to-noise ratios to be obtained by computer averaging a large number of scans.

5.3.5   *Vidicon tube detectors*

The Vidicon tube has been the subject of research as a detector for spectrometry since it, like the diode array, can be used in conjunction with a non-scanning monochromator. It consists of multiple diodes ($15\,000\,\text{mm}^{-2}$) deposited on the face of a cathode ray tube (Fig. 5.5).

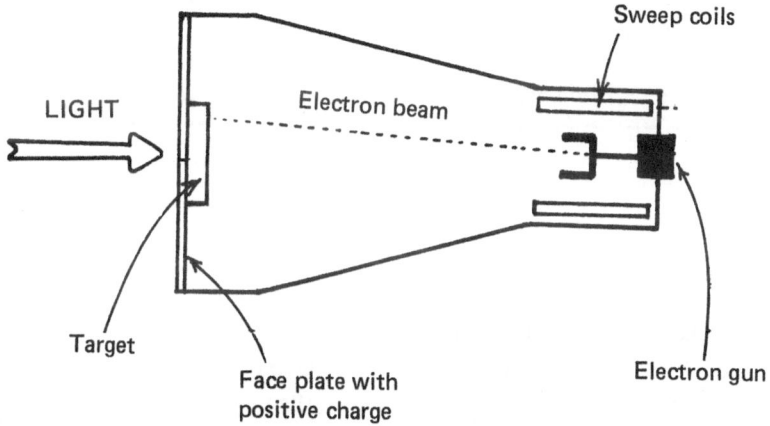

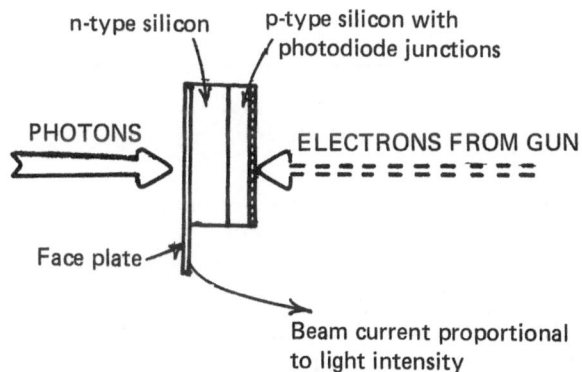

Fig. 5.5 *A Vidicon detector: (a) transverse section (b) section through the target.*

The dispersed radiation from the grating is focused on the outer surface of the window and as the tube is scanned by the electron beam, the beam current increases when it scans a slice of diodes illuminated by a particular wavelength. Thus the current generated in each line scan will be proportional to light intensity and a spectrum may be stored and reconstructed by a computer. This multichannel analyser is experimental at present and is both costly and complex, and since there are only 500 channels available at present, the Vidicon cannot give high wavelength resolution.

5.4   Summary

Table 5.2 summarizes the properties of the different detectors as used in commercial spectrometers the choice being determined by spectral range and sensitivity.

Table 5.2: *Detector summary.*

Type	Range	Sensitivity	Electronics
Barrier layer cell	VIS	Low	Simple
Lead sulphide cell	IR	Moderate	Simple
Silicon diode	UV–VIS	Good	Simple
Phototude	UV–VIS	Good	Simple
Photomultiplier	UV–VIS–near-IR	High	High voltage
Diode array	UV–VIS	High	Complex computer required

General reading

Rabek, J.F. (1982), *Experimental methods in Photochemistry and Photophysics*, John Wiley & Sons, Chichester. pp. 445–529.
Skoog, D.A. and West, D.M. (1971), *Principles of Instrumental Analysis*, Holt, Rinehart and Winston, New York.
Thorn–EMI Electron Tubes (1982), *Photomultiplier Product Catalogue*, Thorn EMI Tubes Ltd, Ruislip, Middlesex, England.

6 Instrument signal processing

This chapter is concerned with the conversion of the small electrical signals, obtained from the detectors described in the previous chapter, into a convenient form for the operator of the instrument to use the data. It is divided into descriptions of the electronic amplifiers required to suit the different types of detectors and methods of signal processing, data processing operations and presentation of data. The different systems are illustrated by examples of specific instruments and, in general, the detailed descriptions are given in relation to these examples rather than under the specific headings.

6.1   Amplifiers

6.1.1   *Pre-amplifiers*

Typically the signals from optical detectors are extremely small (as little as 10^{-11} A for a vacuum phototube). It is essential to amplify such signals to a convenient working level without loss of signal information and without introducing noise. This is normally performed by using a pre-amplifier typically mounted close to the detector itself to avoid the risk of spurious signals being picked up on the intervening leads.

For most types of photo-detector, the signal can be considered as a small current from a very high impedance source and the pre-amplifier is either a current amplifier or a current to voltage converter. By far the most common principle employed is the virtual earth amplifier in which the amplified signal is fed back to the input with reversed polarity in order to maintain the input at essentially constant potential. This type of amplifier is illustrated, in principle, in Fig. 6.1. In an ideal amplifier, the input to the amplifier itself draws no current and, therefore, the whole of the detector signal (current) passes

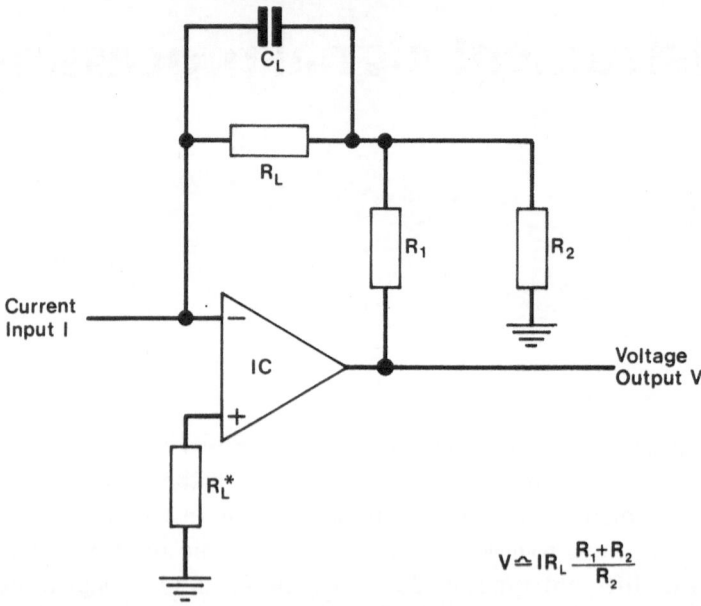

Fig. 6.1 *Basic circuit of a photo-detector pre-amplifier. IC is a FET-input opera-tional amplifier, e.g. Analog Devices AD547LH. Resistor R_L^* is for balancing and may be replaced with a shorting link.*

through the load resistor R_L which is performing the feedback func-tion; thus the amplifier will give a voltage output

$$IR_L(R_1 + R_2)/R_2$$

where I is the detector output current and assuming R_L is large com-pared with R_2. Typical values of R_L range from a few thousand ohms for certain photo-conductive or photo-emissive detectors to about 2000 Mohm for phototubes. As already mentioned, an essential requirement of such amplifiers is that a negligible current is drawn by the input and this condition is well met by modern field effect tran-sistor (FET) input operational amplifiers. However, before such amplifiers were available, it was normal to use valves; for the most demanding applications like phototube signals special valves with exceptionally low grid current – electrometer valves – were used. Fig. 6.2 shows the circuit of an early amplifier employing such valves. Note the use of batteries for power supplies for stability and to avoid mains frequency pick-up. Such amplifiers have now been generally replaced, even in the early instruments which used them, since electrometer valves are no longer manufactured, their function being better performed by FETs.

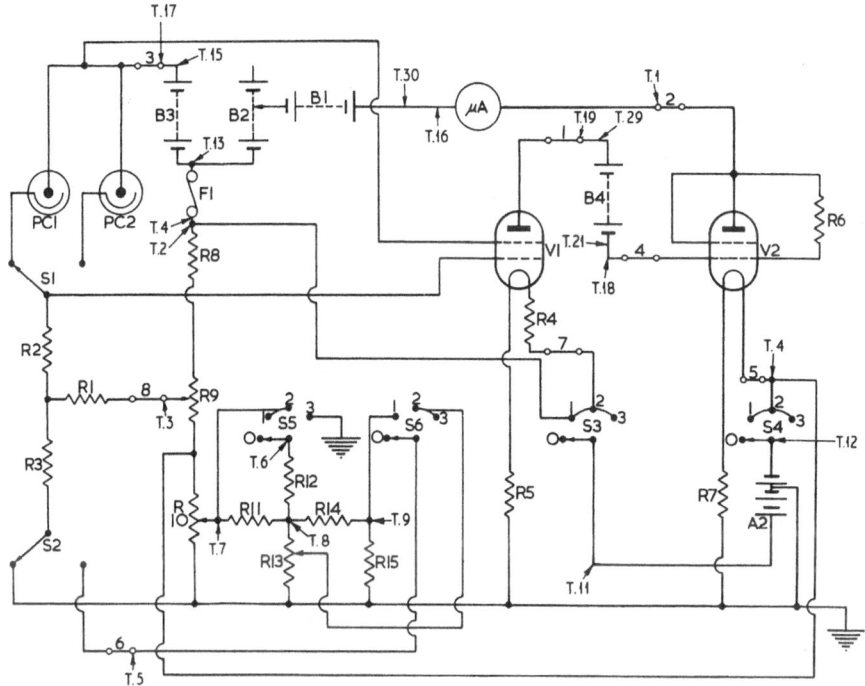

Fig. 6.2 *Circuit of an early manual null phototube amplifier employing electro-meter valves. By courtesy of Pye Unicam Ltd.*

(a) *Offset and drift*

For d.c. instruments, i.e. those in which the signal is not chopped, the amplifier output must remain constant in the absence of signal from the detector. Again, modern operational amplifiers perform very well in this respect. The standing offset may be as small as 0.25 mV and the change of this offset with temperature 1 μV $°C^{-1}$. Before such amplifiers were available, one technique for eliminating offset and drift was to switch both input and output between the signal and earth – the chopped d.c. amplifier. The circuit of such an amplifier is shown in Fig. 6.3. Note that this design was for a relatively low impedance signal from a selenium photocell.

(b) *Noise*

As already mentioned, the amplifier should ideally add no random fluctuations to those already present in the signal from the detector. In practice, it will introduce noise – Johnson Noise – due to the

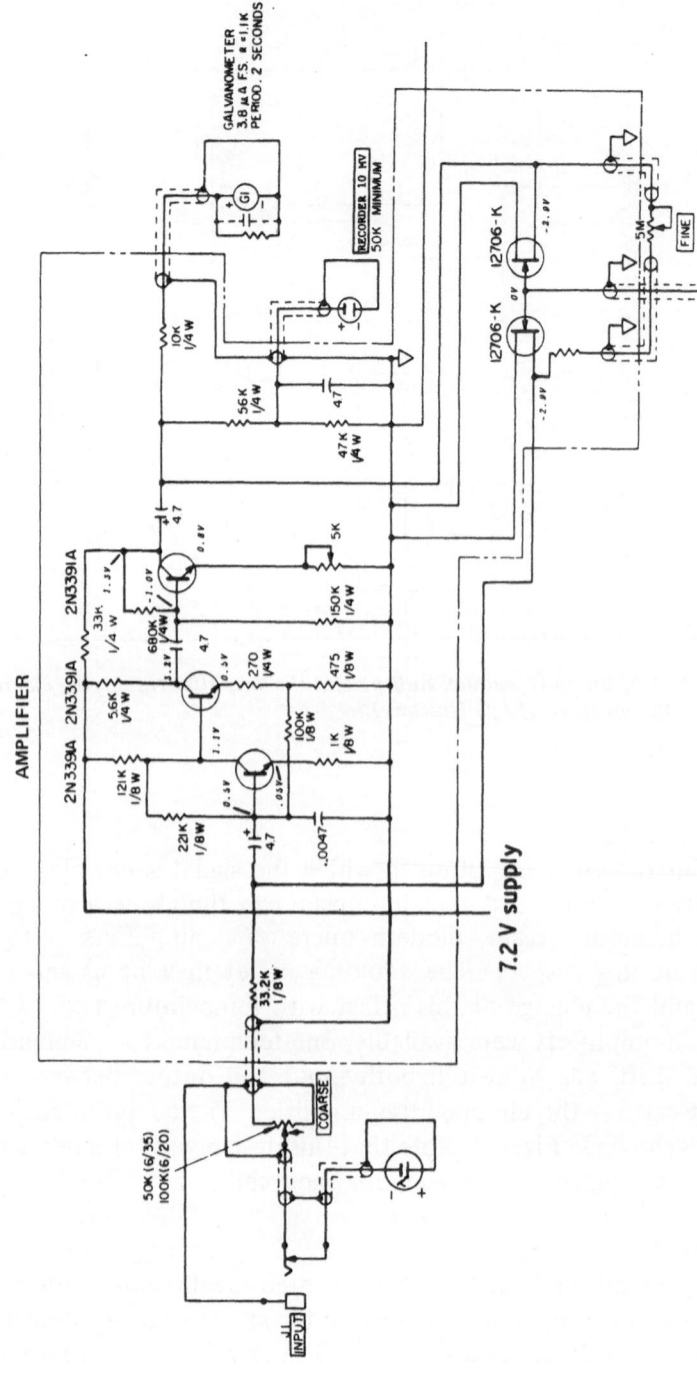

Fig. 6.3 *Circuit of a chopped d.c. photocell amplifier. By courtesy of Perkin–Elmer Ltd.*

inherent behaviour of the load resistor. Such noise is defined by the equation

$$\text{Noise (V RMS)} = (4k\,T R\,\Delta f)^{1/2} \cong 1.3 \times 10^{-10}(R\,\Delta f)^{1/2}$$

where k is Boltzmann's constant, T is temperature (K), R is resistance (ohm), and Δf is the frequency interval. This will typically be the dominant noise source in the case of phototubes, but should not normally be significant for other types of detector. It is normal practice to place a small capacitor C_L in parallel with the load resistor to attenuate noise at frequencies higher than those of interest: the effective frequency f beyond which signal and noise are strongly attenuated is defined by

$$f = \frac{1}{2\pi R_L C_L}$$

In general pre-amplifier design, it is better to err by having this filter capacitor too small rather than too large since further filtering of unwanted high frequency signals will normally be performed later and this initial filtering is primarily to avoid excessive high frequencies interfering with amplifier behaviour, e.g. by causing partial saturation and consequential non-linearity.

6.1.2 *Main amplifiers*

The magnitude of the signal from the pre-amplifier should be such that there should be little risk of introducing noise or spurious signals, e.g. mains pick-up. However, this does not mean that reasonable precautions to avoid such effects should not be taken particularly if, at any point, the signal is significantly attenuated. For the purposes of this discussion, the main amplifier is considered to be that part of the circuit in which other necessary operations on the analogue signal are performed. Such operations are:

(a) Removal of the effect of any standing signal from the detector (dark current) and amplifier offsets.
(b) Filtering out unwanted high-frequency signals and noise.
(c) Conversion from linear to logarithmic scale for presenting the output in absorbance rather than transmittance.
(d) Scaling of the signal to a convenient level.

These functions vary very greatly according to the principle on which the instrument operates. Because there are so many possible variations, the principles will be described in relation to a selection of specific examples.

6.2   Single-beam instruments

6.2.1   *Manual null*

Although instruments of this type are no longer manufactured, a large number are still in use and it is, therefore, relevant to mention the principles involved. Typically, only a single amplifier is involved performing functions of both main and pre-amplifiers. The principle on which such an instrument operates is to exactly balance the signal from the detector with a signal generated from a manually adjusted potentiometer. It can be considered similar in principle to the virtual earth amplifier referred to earlier, only in this case the re-adjustment of the amplifier input to earth potential is performed manually, the amplifier serving to indicate when this condition has been achieved. The circuit of such an instrument is shown in Fig. 6.2. The operation consists, first of all, of setting the amplifier output to zero – null – using the dark current potentiometer R9 with no light falling on the detector. Subsequently, with the beam open and solvent or reference in place, the slitwidths or the sensitivity potentiometer R10 are again adjusted to null with the slide wire potentiometer R13 and its 1.0 T (zero A) position. Then the sample is introduced and a further null obtained using R13. The reading on the potentiometer is then a direct reading of transmittance on the linear scale or absorbance on the non-linear scale. The advantages of such a system are that a very direct measure of the detector photo-current is obtained which is independent of the characteristics of the amplifier although, obviously, any drift of the amplifier which changes the null point will make readings difficult.

6.2.2   *Meter readout*

The circuit of a very simple meter readout amplifier is shown in Fig. 6.4. Again, a single amplifier performs functions of both main and pre-amplifiers. In this instance the detector is a selenium photo-cell and the fact that this is a comparatively low impedance device permits both fine and coarse scaling, i.e. gain adjustment, to be

---

Fig. 6.4 *Circuit of a simple photocell amplifier for meter readout. By courtesy of Perkin–Elmer Ltd.*

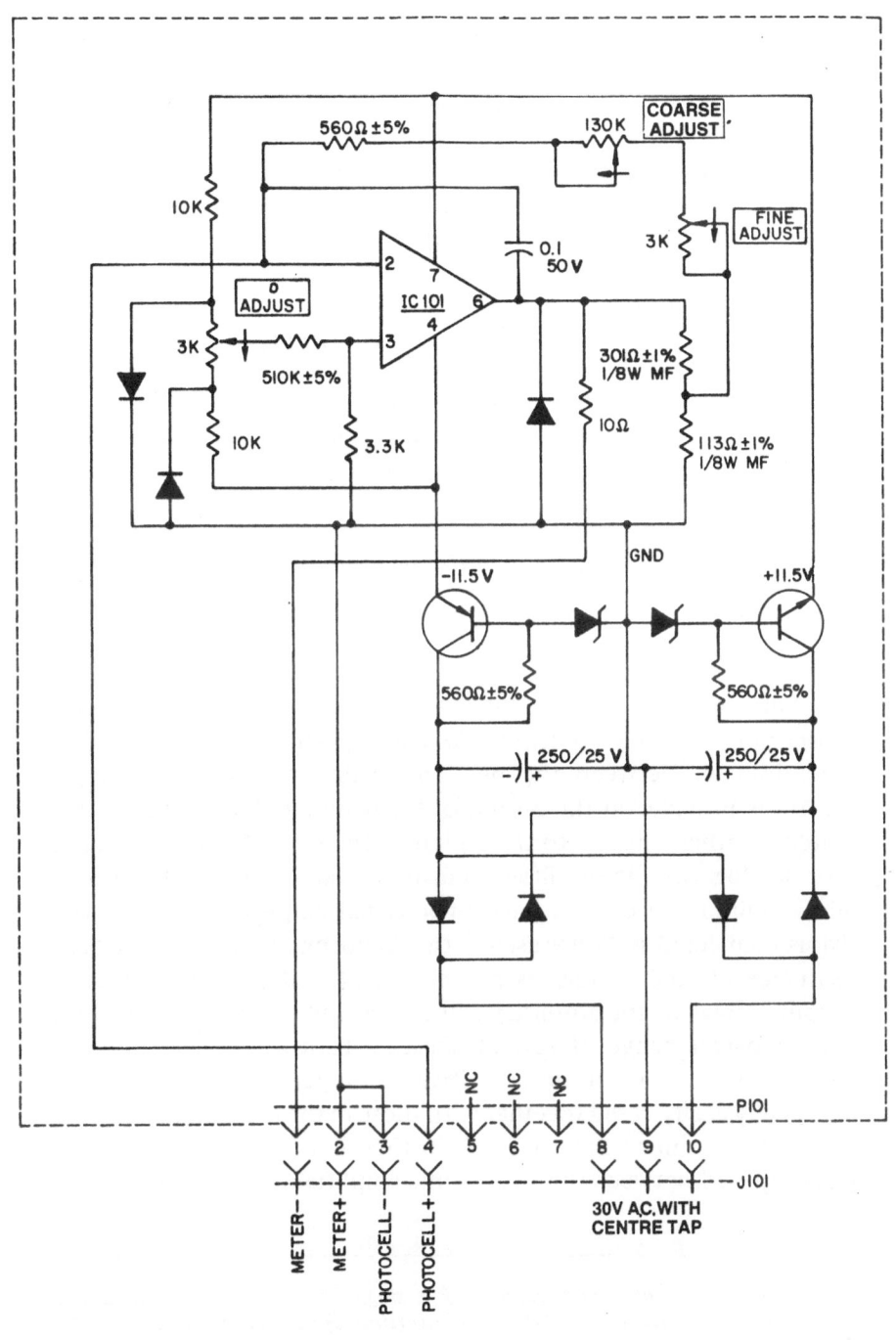

performed by alteration of the load resistor which includes the 'coarse' and 'fine adjust' potentiometers. Balancing of the dark current is performed by adjustment of the potential at the reference (non-inverting) input to the operational amplifier IC101 using the 3 kohm '0 adjust' potentiometer. In order to increase the gain, the feedback resistor is connected not directly to the output of IC101 but to a point on a potential divider between this and earth. The readout meter is connected directly between the amplifier output and earth. The operation of such an instrument is, first of all, to set the meter to zero with the beam blanked using the '0 adjust', then to set the meter to 1.0 T with the reference or solvent in place using the coarse and fine adjustments. After this the transmittance of any sample can be read directly from the meter. An interesting feature of this simple design is that the time constant of the noise filter, comprising a 0.1 μF capacitor and the load resistor, increases as the gain increases, thus limiting the increase in noise.

6.2.3   *Digital readout with logarithmic amplifier*

Fig. 6.5 shows the analogue part of the circuit of an instrument employing a vacuum phototube detector and incorporating a logarithmic amplifier for absorbance readout. The pre-amplifier is of conventional design with a 1000 Mohm load resistor and dark current balance at the non-inverting input. Coarse gain control is by means of the potentiometer between the pre- and main amplifiers. A fine gain control is provided at the output of the main amplifier, which can be switched either directly to the digital voltmeter (DVM) for display in T or via a logarithmic amplifier for display in absorbance. The logarithmic amplifier is a conventional operational amplifier with two transistors connected as diodes used as the feedback resistor. The effective resistance of such diodes as a function of voltage is such that the output voltage of the amplifier will vary as the logarithm of the input voltage over a range of several decades. However, since this diode characteristic is temperature sensitive, a compensating amplifier using diodes physically cased together with those in the logarithmic amplifier is used to minimize the effect. A further variable gain amplifier can be interposed between the logarithmic amplifier and the display input

---

Fig. 6.5 *Complete analogue circuit of a single-beam instrument including a logarithmic amplifier for absorbance conversion. By courtesy of Perkin–Elmer Ltd.*

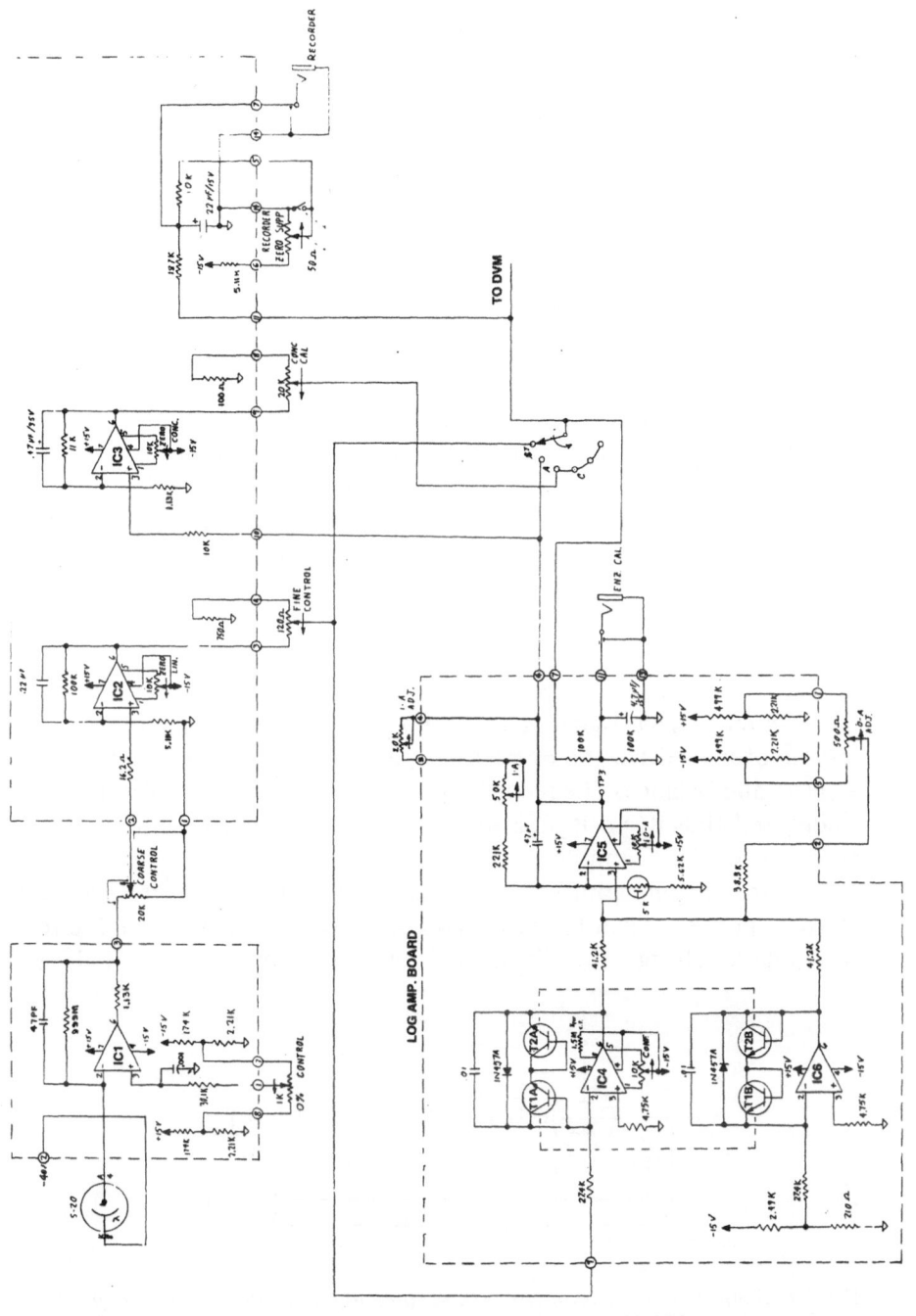

in order to scale the absorbance reading so as to obtain a direct reading in concentration of the sample or other arbitrary units.

6.3   Double-beam instruments

6.3.1   *Optical null*

Although few commercial UV models have been produced using the optical null principle, a description of the amplifier systems is given here for completeness. In an optical null double-beam instrument, the light intensity passing along the reference beam path is attenuated with a mechanically variable attenuator (optical wedge) to match the intensity passing along the sample beam path. Light from the mono-chromator is directed by a 'chopper' alternately along the two paths to the detector and, therefore, in the null condition the detector experiences no variation in signal as the beams are switched. An off—null condition results in a signal alternating at the chopper (beam switch) frequency and it is this signal which is amplified by an a.c. coupled amplifier. Since any dark current or amplifier offset is not modulated, it does not contribute to the a.c. component of the signal and can be ignored. The a.c. signal is rectified by switching synchronously with the optical chopper so that the sign, positive or negative, of the output will depend on which beam, sample or reference, is the more intense. This signal is then applied to a further amplifier and servo-motor to drive the optical wedge in the necessary direction to achieve null balance. The recorder pen is mechanically coupled to the wedge and thus its position is directly related to the transmittance or absorbance of the sample.

Automatic gain control (AGC) is employed to keep the wedge servo response constant, irrespective of changes of signal level due to wavelength, changes in slit width, or sample attentuation. This is

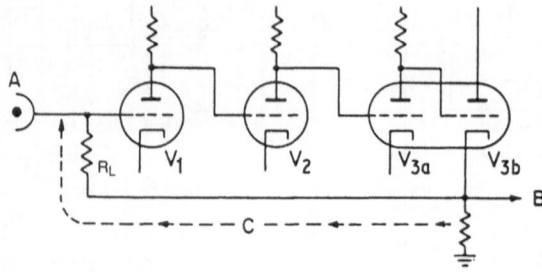

Fig. 6.6 *Simplified schematic of a valve pre-amplifier for use with a photo-multiplier.* $R_L = 100$ *Mohm.*

achieved by varying the high voltage applied to the photomultiplier so as to maintain at a constant level a higher frequency signal generated by interruption of the beam by a mask on the optical chopper. It should be noted that the wedge in instruments of this type is logarithmic, i.e. the position of the recorder pen which is coupled to the wedge is linear in absorbance. The combination of AGC with a logarithmic wedge maintains the servo response constant at all wedge positions which would not be true in the case of a linear wedge.

6.3.2 *Ratio recording systems*

(a) *Two-detector instruments*

Some instruments employ separate detectors for the sample and reference beams. The beam from the monochromator is interrupted by a synchronous chopper and then passively split into the sample and reference paths. The electronic system of one of these instruments that has some unusual features will be described in some detail.

Separate, fairly conventional, virtual earth, valve pre-amplifiers with 100 Mohm load resistors are used for the photomultiplier detectors (Fig. 6.6). The signal from the reference channel is maintained constant by a servo system operating on the slitwidths. The common high voltage applied to the two photomultipliers is adjusted manually and, as a result of the slit servo, has the effect of varying the instrument resolution.

A very interesting feature of the electronic design is that there is no demodulation of the a.c. signals. Ratioing of the sample signal to the reference signal is performed using the slide wire of the potentiometric recorder (Fig. 6.7). The a.c. reference signal is applied across the slide wire (C) and the output from the potentiometer wiper (D) compared with the a.c. sample signal. The difference between these two signals is amplified and fed to the control winding of the pen

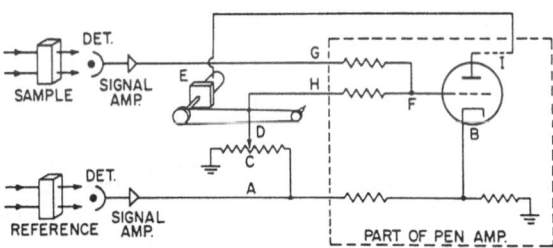

Fig. 6.7 *Simplified schematic of the ratioing system of an early two detector double-beam instrument.*

servo-motor (E), and the phase of this difference signal depends on whether the signal from the potentiometer is greater or less than the sample signal. The reference winding of the servo-motor is fed from mains (the optical chopper operates synchronously at mains frequency) and, therefore the servo-motor is driven in the direction to achieve recorder balance. As in the previous example, the dark current and amplifier offsets are not modulated and, therefore, have no effect on the achievement of balance. It should be noted that in both the above examples, AGC serves only to maintain servo response constant and is not critical to correct measurement of transmittance or absorbance values. Similarly, changes of gain of the servo-amplifiers are not critical; however, in the latter example the gain of the separate pre-amplifiers prior to the comparison in the potentiometric recorder does need to be stable. One disadvantage of this type of system is that the provision of any form of filtering (low-pass filtering to eliminate unwanted high frequencies) is very difficult. It can be performed to some extent, however, by providing excessive damping on the recorder – see Section 6.5.

These instruments incorporate an electromechanical system for baseline correction, Fig. 6.8. This takes the form of a multitapped

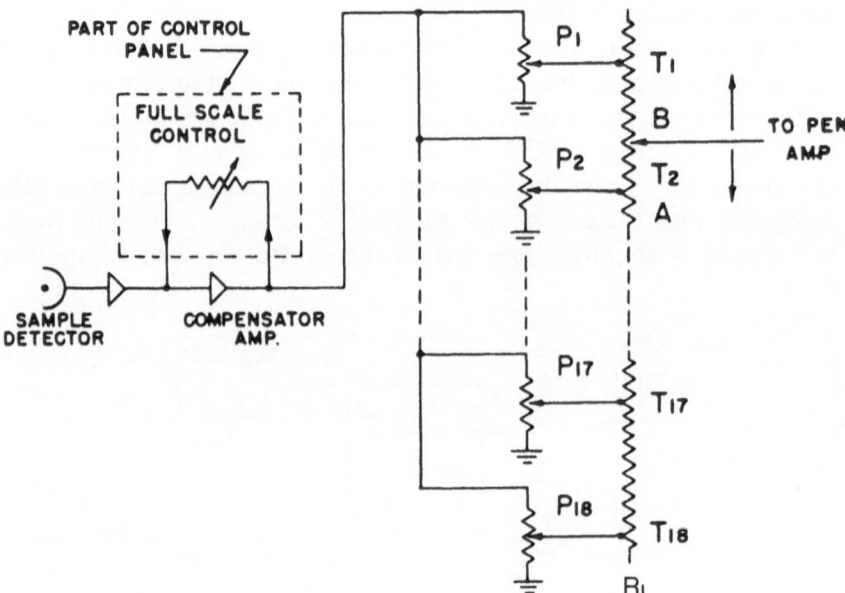

Fig. 6.8 *Circuit of an early electromechanical baseline correction system.*

potentiometer (A) the wiper of which (B) is coupled to the mono-chromator wavelength drive. Each of the taps goes to one of the trimmer potentiometers P_1 to P_{18} which act as pre-set attenuators of the sample signal. The system is set up manually and results in the baseline being corrected at each of the 18 points corresponding to the tap positions, T_1 to T_{18}, with linear interpolation between.

(b) *Single detector – non AGC dependent*
In this type of instrument a single detector measures both sample and reference channels in sequence, together with a dark period. The required signals are the difference between the sample and dark signals and the reference and dark signals, respectively. Various methods have been employed for obtaining these differences, and one form will be described.

In these instruments the signal from the conventional pre-amplifier (10 Mohm load to suit a photomultiplier detector) is fed to the main amplifier through a capacitor, the amplifier side of which is earthed by an FET switch during the dark period. This results in the voltage across the capacitor corresponding to the dark current output from the pre-amplifier. Since the main amplifier input draws negligible current, the voltage across the capacitor remains constant during the subsequent measurement of the sample and reference signals and, therefore, in effect subtracts the dark signal from these. Following the input amplification stage of the main amplifier, the signal is switched during the respective sample and reference periods through FET switches to separate sample and reference integrating circuits. Each of these produces, at the end of the appropriate measuring period, a voltage level corresponding to the integral of the respective sample or reference signal corrected for dark current. During the next dark period the ratio of these two voltages is generated, after which the two integrators are reset to zero by further FET switches. In principle, the ratioing is performed by measuring the time required to integrate a current corresponding to the voltage output from the reference integrator until it reaches a level equal to the signal integrator output voltage. This time is, again, converted to a voltage by the integration of a fixed current for that time duration.

The complete sequence just described is performed once per complete cycle of the optical chopper, which is so arranged that each integration time corresponds exactly to one mains frequency cycle. (*Note*: the object of this is to ensure that any mains frequency pick-up integrates exactly to zero.) The remainder of the system is essentially the same as that described in Section 6.2.3, i.e. the scaling facilities, logarithmic amplifier for conversion to absorbance and digital voltmeter for display. However, the signal is also fed to a recorder output. The instrument does incorporate automatic gain control through the high voltage applied to the photomultiplier, this voltage being electronically servoed to maintain the signal in the reference channel at a constant level. However, the system does not depend critically on the behaviour of the AGC since the electronic ratioing will operate satisfactorily over a fairly wide range of reference levels. Since a common amplifier is used for both sample and reference signals, the gain stability is not important. However, the linearity of the amplifier is important since any departure from linear behaviour would affect the linearity of the transmittance measurement.

(c)   *Single detector – AGC dependent*

The principle employed in this type of system is similar to that described above except that a highly accurate electronic servo system is employed to control the high voltage to the photomultiplier so that the reference signal level remains exactly constant. When this is done, it is not necessary to measure the ratio of the sample and reference signals since the sample signal itself represents that ratio when the reference signal is constant. The main features of a typical circuit are shown in Fig. 6.9. The input from the conventional pre-amplifier is fed via a capacitor to the input of the main amplifier which is earthed during the dark cycle. Following the first stage of the main amplifier, the signal, minus a constant comparison voltage, is switched during the reference period to an integrator, the output level of which provides the reference voltage for the photomultiplier high voltage supply. When the reference signal is equal to the comparison voltage, there is no net input to the integrator. However, if the reference signal is lower than the comparison voltage, the output of the integrator will drop resulting in a higher photomultiplier voltage (a voltage inverter is interposed between the integrator and the output to the high voltage supply). Conversely, if the reference signal is too high, the high voltage to the photomultiplier will be reduced. In this instrument, the signal during the sample period is fed not to an integrator but to a low-pass filter circuit after which, like the instrument described above, it is fed to a digital voltmeter either directly or through a logarithmic amplifier. The frequency of the optical chopper is not synchronous with mains, but is deliberately chosen to be well away from mains frequency or any harmonic of it.

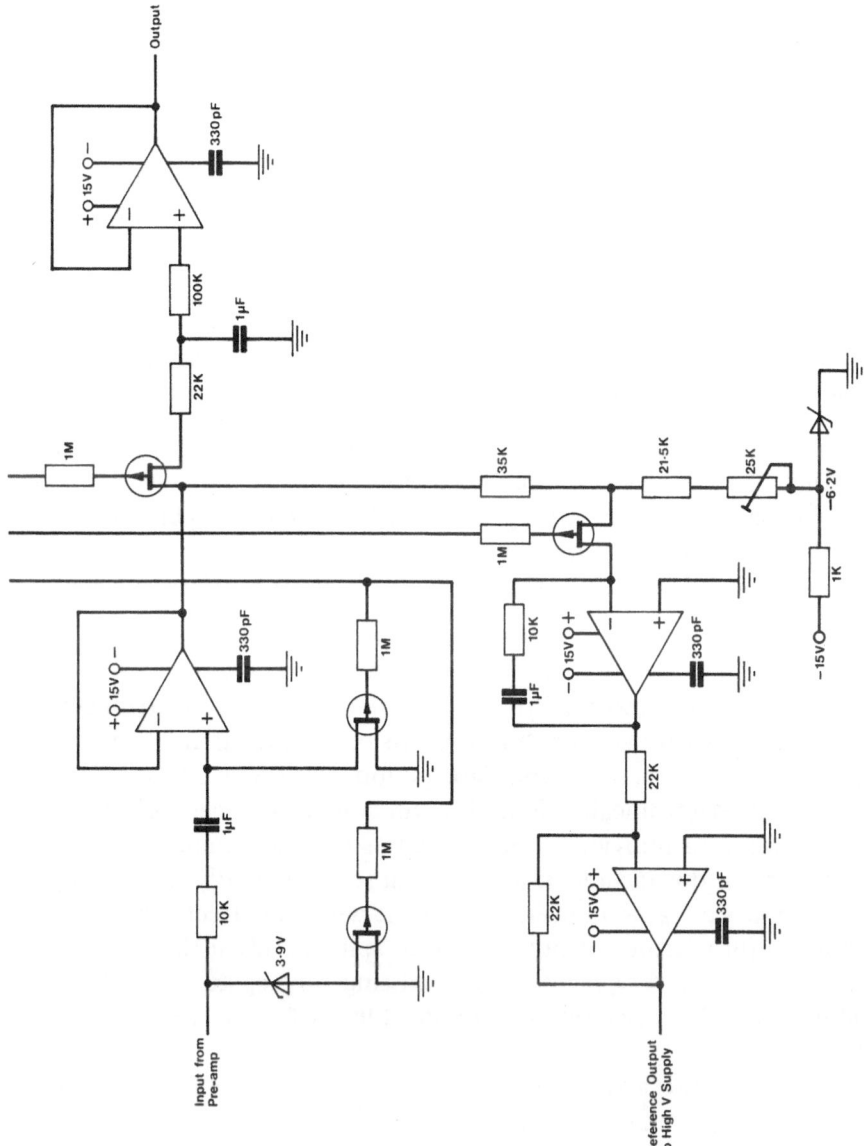

Fig. 6.9 Main features of the circuit of an AGC dependent double-beam instrument used in the Perkin-Elmer Model 550. By courtesy of Perkin-Elmer Ltd.

This is an alternative approach to minimizing the effects of mains frequency pick-up to that described above of integrating for exactly one mains frequency cycle.

6.4   Microprocessor-based instruments

6.4.1   *Analogue-to-digital conversion*

The analogue-to-digital converters used in the digital readout instruments mentioned in Section 6.2.3 were not described in detail since they are essentially standard digital voltmeters. The design of such devices has been very fully covered elsewhere. However, the special requirements for digitizing instrument signals, and in particular UV instruments, warrant some discussion.

If an analogue signal is to be digitized without loss of information, three conditions must be met:

1. The smallest measurable increment must be less than the noise (half the RMS noise is usually adequate).
2. The sampling rate must give at least two samples per cycle of the highest *noise* frequency.
3. The dynamic range must be sufficient to accept the largest signal.

Conditions (1) and (3) are reasonably obvious, but condition (2) requires further explanation. Consider, for example, the low light level signal from a photomultiplier used in double-beam instruments where, for signal processing requirements one digitization is required for each sample, reference and dark period, say every 20 ms. If the signal is not electronically filtered it will consist of a mass of 'spikes' corresponding to individual photon counts (the randomness of these is the noise in this instance). If this signal is momentarily sampled by the analogue-to-digital converter (ADC) it will read effectively 0 or 1 depending on whether or not it reads a 'spike'. Only if the sampling rate is sufficient to guarantee not missing any spike and all the readings over the measuring period co-added will there be no loss of information.

Rather than use a very high sampling rate, it is generally better to filter out the high noise frequencies and this can be done either by (a) using a 'low-pass' filter, or (b) by integrating the signal over each measuring period. When method (a) is used care must be taken to ensure that the signal frequencies of interest are not also filtered; this is particularly important for double-beam instruments for which method (b) is generally preferred.

In a UV instrument using a photomultiplier detector, the noise will normally increase as the square root of the signal. Thus, although the total dynamic range (ratio of largest to smallest signals) may be very large — up to 10^6 :1 — the signal-to-noise ratio may never exceed 10 000:1 because of the larger noise at large signals. This situation can be used to advantage to avoid the need for very high precision — and very expensive — ADCs by using, say, a 12 or 14 bit ADC with 'gain switching'. In such a system the gain of an amplifier, preceding the ADC, is switched in steps of a factor of 2 under microprocessor control to bring the signal to a convenient level for digitization. This arrangement still complies with conditions (1) and (3) above.

Details of the many types of ADC will not be given here, although one unusual design is described in the next section. Readers are referred to the many publications on the subject and, in particular, to the paper by Owens [1].

6.4.2   *Ratioing by microprocessor*

Microprocessors are now used in many commercial instruments for processing all the signals generated in the basic types of single-beam or double-beam ratio recording instruments described above. In addition, the signal can be digitized immediately after the pre-amplifier and the functions of the correction for dark current and ratioing performed digitally.

The pre-amplifier of a typical instrument together with the analogue-to-digital converter is illustrated in Fig. 6.10 and is a true current amplifier with a current gain of 50. The signal from the current amplifier is fed to an unusual form of ADC which operates under the control of the microprocessor. The current from the pre-amplifier is integrated by ICA02 until the output of the integrator triggers the comparator ICA03. This causes the integrator to be discharged by an accurately known amount by causing an accurately defined current to flow out for a precisely measured time. The number of times during a measuring period that this 'charge dump' occurs is an exact measure of the integrated input current. Although the discharging operation is performed very quickly, the maximum number of discharges possible in a typical measuring period (1024) is insufficient to give a highly accurate reading of the signal. Therefore, at the end of the measuring period, the residual fraction of a charge dump remaining in the integrator is measured by a comparison with the signal generated by ICA04, a digital-to-analogue converter, which achieves

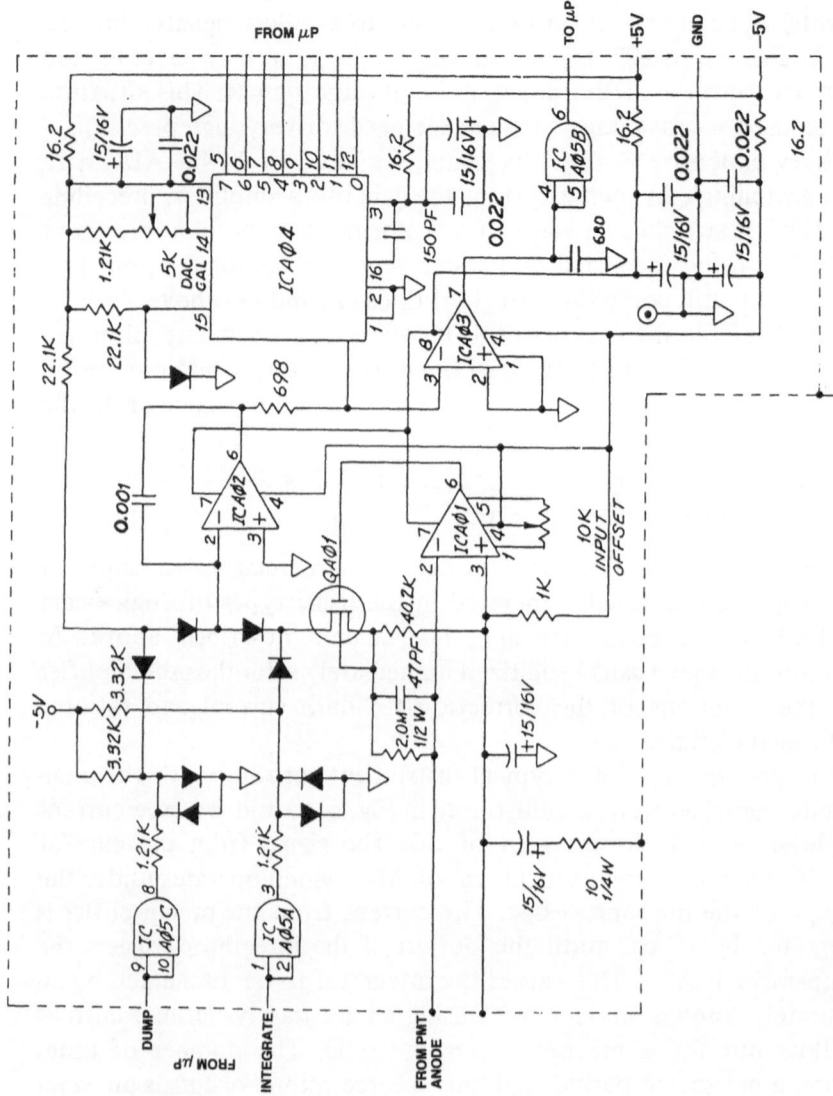

Fig. 6.10 Circuit of the pre-amplifier and ADC of a microprocessor-based instrument used in the Perkin–Elmer Model 552. By courtesy of Perkin–Elmer Ltd.

the balance by successive approximation measurements under the control of the microprocessor. During this successive approximation measurement, which takes less than 1 ms, no integration is performed. The full range of this ADC is 18 bits. An analogue-to-digital conversion is performed during each dark, sample and reference period and the results stored in the microprocessor. The measured dark level is digitally subtracted from the sample and reference signals, respectively, and the corrected sample signal is then divided by the corrected reference signal. Subsequent operations are common to all the microprocessor-based instruments and are discussed together in the following section.

6.4.3 *Microprocessor signal processing*

Having obtained the ratio of sample and reference signals in a ratio recording double-beam instrument, there are no differences in principle between the signal (data) processing for a single-beam or double-beam instrument in so far as fixed wavelength measurements are concerned. These common functions are discussed first and the special functions associated with scanning are discussed in the following section.

(a) *Non-scanning functions*

The main data processing operations for fixed wavelength measurements are auto zero, i.e. the automatic setting, on command, of the instrument readout to zero absorbance or 1.0 T; conversion from transmittance to absorbance when this has not been performed in the analogue circuitry; scaling of absorbance to give a readout in concentration or other arbitrary units and integration. In addition, it is possible to perform various kinetic measurements. There is one additional function which is usefully performed in non-chopped single-beam instruments and that is the automatic setting of zero T, i.e. dark current correction. This is performed by introducing a shutter into the beam under the control of the microprocessor, measuring the dark current and then subtracting the digital value of the dark current from all subsequent measurements. The requirements for auto zero and concentration scaling are so simple that they will not be discussed further, but absorbance conversion and integration warrant some further description. In order to save time within the microprocessor, absorbance conversion is normally performed, not by calculation from first principles, i.e. from the mathematical expansion for a logarithm, but by use of pre-calculated values – a look-up table – with interpolation between successive points. Invariably, all data handling operations are performed in binary arithmetic (usually integer but sometimes floating point) and the results are only converted to decimal format for display purposes. This means that logarithms are calculated to base 2 (although the values in the look-up table can be already scaled to base 10) and, therefore, the look-up table need only cover a range of signal from 1 to 2. This means that very few points are necessary in the look-up table to achieve the necessary accuracy of $\pm 10^{-4}$ absorbance units. A well-written program for a typical 8 bit microprocessor will perform a transmittance to absorbance conversion using this technique in the order of 1 ms.

When making individual measurements at a fixed wavelength, by far the best method for minimizing noise is to integrate the signal for as long as is convenient. Integration, which is simply the summation of successive measurements or digitizations of the signal followed by division by the number of measurements summed, can be performed very readily with the microprocessor. For ultimate simplicity in performing the division, it is common to restrict the number of measurements being summed to powers of two.

(b) *Requirements for scanning instruments*
In contrast to single-point measurements, the best method of noise filtering for a scanning instrument is definitely *not* integration since this will always have the effect of attenuating peak heights. One of the best noise filter systems for scanning conditions is that proposed by Savitzky and Golay [2]. They described a series of filter functions of which the quadratic–cubic is the one normally chosen for instrument use. This is slightly inferior in performance to the higher order filters but is much simpler to use. It has the property that, so long as the true data are only varying as a quadratic function of time or wavelength, they will be undistorted by the filter. This filter function can take significant calculation time unless special programming tricks are employed, but it is now possible to perform such filtering over an arbitrary number of points at a rate of approximately 1 point every 3 ms using an 8-bit microprocessor. The typical data flow diagram for the data processing in a hypothetical double-beam scanning instrument is shown in Fig. 6.11. Note that the filtering for the digital display and the recorder output are different, with the display using a 'moving block average', i.e. a continuously shifting integration. Other functions which are specific to scanning instruments are the

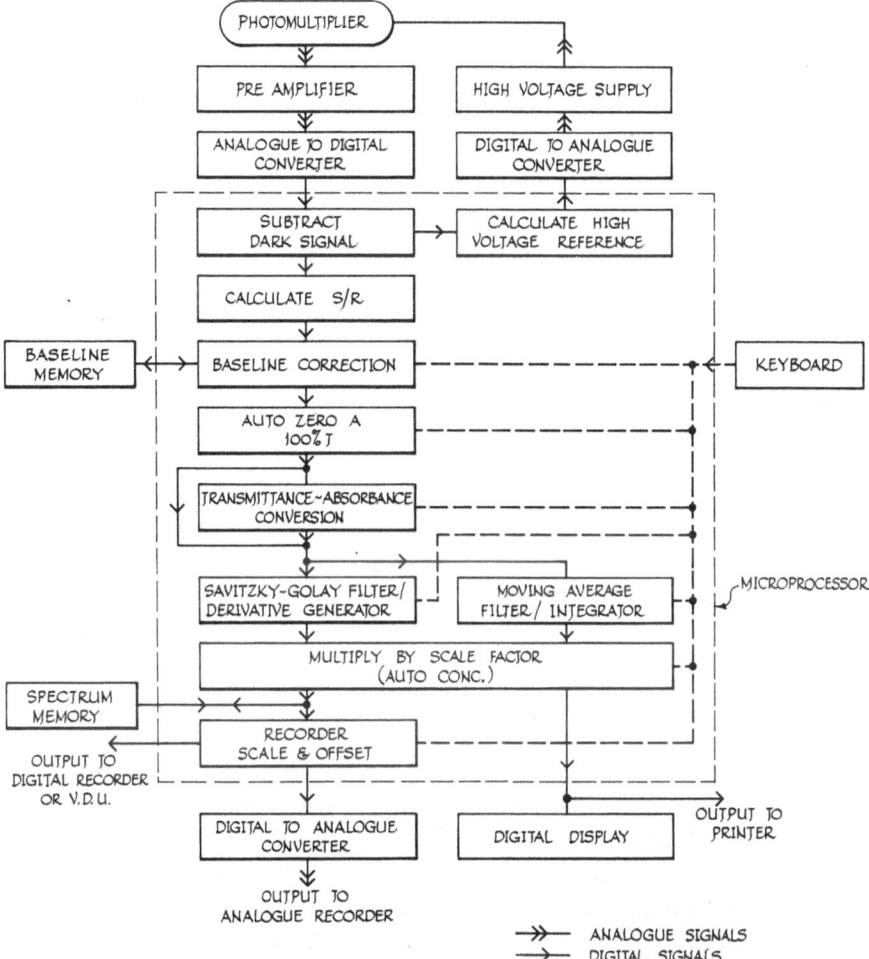

Fig. 6.11 *Simplified signal processing sequence in a hypothetical microprocessor-based instrument.*

generation of first, second and higher derivatives. These functions can be generated readily by substituting the Savitzky–Golay quadratic smoothing filter with one of the other appropriate procedures described in their paper. The use of derivatives is discussed further in Chapter 11.

Finally, it is very easy with a microprocessor to provide background correction. In principle, all that is necessary is to record in the microprocessor memory the spectrum with the sample position empty or with a reference or solvent in position, and then to subsequently ratio the measured spectrum against this background. This facility is now available in the majority of instruments. In principle, this technique can be used to convert a scanning single-beam instrument into, effectively, a double-beam instrument, although it should be noted that the long-term stability requirements for source and amplifier, which are unnecessary in double-beam instruments, again become critical. There are a large number of other data processing operations which could, in principle, be performed within an instrument but, at present, are normally carried out only using external computers or data stations. No doubt more of these will be incorporated within instruments in the future.

6.4.4 *Use of multiple detectors*

'Multiple' in this sense refers to detector systems capable of simultaneously making multiple independent measurements of the optical signal, usually over a range of wavelengths but sometimes for spatial discrimination of the sample at a single wavelength. The main types in use are TV camera tubes (vidicons, plumbicons, etc.) and self-scanned diode arrays. A full description of the electronic systems involved is beyond the scope of this book and readers are referred to the device manufacturer's literature for full details; however, some general comments in relation to their use in spectrometers are relevant.

Irrespective of detector type, the outputs from the individual elements are 'scanned' in time rather than measured strictly simultaneously. Thus, in order not to 'waste' the signal during the high proportion of time that each element is not sampled, it is essential that the element acts as an integrator of the signal for the time between sampling. The maximum charge (typically of the order of 10 pC) which can be stored as the integral in each detector element sets an upper limit on the total signal x time product. Thus the largest signal which can be measured varies roughly as the inverse of the interval between scans of the array. The ratio of the largest signal to the dark current noise determines the dynamic range which is usually limited to about 10^4 to 1, very much smaller than the corresponding figure for a photomultiplier.

The use of high sampling rates (say 200 kHz for a 1024 element array, i.e. each element sampled approximately once every 5 ms) eases this problem but puts difficult requirements on the rest of the system: thus, for a microprocessor system – almost a foregone requirement – the signal would have to be digitized and added into memory once every 5 μs in the above case.

The data-handling system must take account of the fact that the sensitivity of individual detector elements may vary, both in absolute terms and with wavelength. Three methods which have been used in this context are:

(a) Storing in memory the relative response of each element and using these data to scale the output signals.
(b) In double-beam instruments the ratioing of the output of each element for 'reference' and 'sample', respectively.
(c) The co-adding of the signals from each element in turn as the spectrum is scanned across the array, as described by Hicks *et al.* [3].

An interesting paper on the use of multiple detectors is that by Torr and Devlin [4].

6.4.5 *Use of visual display units*

The use of visual display units (VDUs) for the display of instrument parameters and spectral data is increasing and a very brief description of the method of use is given here; for more details readers should refer to the literature of the display manufacturers and the makers of the specialized control integrated circuits.

Almost all VDUs at present used for this purpose are of the 'raster scan' type, similar to a normal TV set. Data are displayed by modulating the electron beam as it scans (usually only 'on' or 'off' but sometimes with intermediate intensities) so as to put 'dots' on the screen in the positions necessary to generate the required pattern. The timing of the modulation is normally controlled by special purpose integrated circuits, including a character generator if text is to be displayed and a graphics controller if data, e.g. spectra are to be displayed in graphical form. When a graphics system is employed, it is usually necessary to have a screen memory with one (or more) bit per pixel, that is each possible dot position on the screen. For a typical screen with a resolution of, say, 512 by 256 this would require 16K bytes of memory. It should be noted that a spectrum of, say, 1024 data points can be stored in 2K bytes of memory with an ordinate precision of 1 part in 64 000 whereas, once it has been transferred to the screen memory for display, the ordinate precision is reduced to 1 part in 256 and half the data points will have been discarded. For this reason data should always be retained in memory or recorded elsewhere in addition to being displayed on the screen.

6.5   **Recorders**

Unlike IR spectrometers, UV instruments do not typically have built-in recorders and often a wide choice of recorders is available. It is important to consider the characteristics of recorders in relation to the data being plotted and the accuracy of the representation of those data. The standard type of recorder which has been in use for many years is the potentiometric strip chart recorder. These operate by using a servo to drive the wiper of a potentiometer to generate a voltage which exactly balances the input voltage, the recorder pen being coupled to the potentiometer wiper by some mechanical system. There are a number of inherent characteristics of such recorders which should be considered.

6.5.1   *Dead band*

Some level of power is always necessary to overcome friction in any mechanical system and cause movement. This is true of recorders and the result is that there always has to be a finite degree of imbalance between input voltage and that generated by the wiper of the potentiometer before the servo-motor will start to correct the imbalance. This error is referred to as the 'dead band' and is typically of the order of 0.2% of full scale. In principle, the dead band can be reduced by increasing the gain of the recorder servo amplifier, but ultimately this will cause the servo to overshoot the correct balance condition and go into oscillation.

6.5.2   *Velocity feedback*

To minimize the risk of such oscillation, most recorders employ velocity feedback, which means that movement of the servo will generate a signal proportional to the rate of that movement and tending to stop the movement. In many recorders both the servo amplifier gain and the amount of velocity feedback are adjustable. The ideal setting for the velocity feedback is the condition known as 'critically damped'. This is best described as the value which is just sufficient to prevent the recorder overshooting the balance position. Further increase of the velocity feedback will only slow the response of the recorder and serves no useful purpose unless it is deliberately used as a means of additional filtering as mentioned in Section 6.3.2.

6.5.3   *Slew rate*

The design of the servo system in most recorders is such that there is a maximum speed at which the pen will move across the chart irrespective of the magnitude of the imbalance. This is referred to as the slew rate. It will often be referred to in terms of the time for the pen to move across full scale (or 98% of full scale) with a step function input. It should be remembered that if a recorder takes half a second to reach full scale, it does not mean it will adequately follow a full scale triangular peak of one second full width. The reason for this is that the imbalance has to be a significant proportion of full scale before the pen reaches slewing speed. There is no simple rule for ensuring that a recorder is capable of accurately plotting a spectrum at a given scanning speed other than by observing the effect of changing the scanning speed. However, it may be assumed that if the half-height width of a peak is more than twice the full scale recorder response time, any residual error will be small.

6.5.4 *Linearity*

Although recorder potentiometers are made very precisely, they are still not perfect and errors in linearity of up to $\pm\frac{1}{2}\%$ are not uncommon. This effect, which will normally be reproducible but of arbitrary shape, and the effect of dead band, which will normally reduce peak heights, should always be considered when making accurate measurements from recorder traces.

6.5.5 *Input impedance*

Recorders will be rated for a maximum supply impedance and, equally, the recorder output of instruments may be rated for a minimum recorder input impedance. The effect of too high a supply impedance to a recorder will be to make the response sluggish and possibly to upset the linearity, and too low a recorder input impedance will have a similar effect. In many recorders the input impedance becomes infinite at balance. If this is the case then linearity will not be affected by impedance mismatch but the sluggish effect may still be observed.

6.5.6 *Stepper motor driven recorders*

Very recently, recorders have become available which do not use a potentiometer but rely on counting the steps of a stepper motor or position encoder to keep a digital record of the pen position. This record is compared with either an incoming digital signal or a digitally converted analogue signal and the error used to drive the stepper motor and the coupled pen to the balance position. Such systems are inherently linear apart from mechanical errors, and the dead band is limited to ±1 step of the stepper motor – usually equivalent to 0.1% or less of full scale. In addition, some recorders with a digital input have a buffer memory in which data are temporarily stored if the slew rate of the recorder is insufficient to follow the data as they come into the recorder. As soon as the rate of change of the input data slows down sufficiently, the recorder will catch up and empty the buffer memory. Such a recorder will always accurately follow input data, no matter how narrow peaks may be, so long as the buffer memory is never completely filled. Typically, some warning signal will be given if this condition has been reached.

6.6   Updating early instruments

With comparatively little work, it would be possible to convert early single-beam instruments to convenient digital readout. A suggested way of performing this is to use an amplifier of design similar to that illustrated in Fig. 6.1 coupled to a standard commercial DVM. The specific values of the load resistor R_L and smoothing capacitor C_L should be adjusted to suit the specific type of detector being used. For instance, appropriate values in the case of a phototube would be 1000 Mohm and 100 pF and for a photomultiplier would be 10 Mohm and 10 nF. The output would be fed directly to the input of the digital voltmeter through a suitable arrangement of coarse and fine adjustment potentiometers for conveniently setting full scale. Many commercial digital voltmeters have a zeroing button which would permit instant correction for dark current. However, it should be noted that, since this correction is made after the coarse and fine scale adjustments, any significant alteration of these will require the dark current offset to be reset. Another useful feature of some DVMs is a readout in decibels (dB). This is a logarithmic scale on which a factor of 10 in signal amplitude gives a value on the readout of 20. Thus, if the decibel readout is zeroed on a reference or solvent, it will subsequently read minus 20 times the absorbance value of the sample. For anyone wishing to go to a little more trouble, a logarithmic amplifier similar to that illustrated in Fig. 6.5 could be constructed and introduced between the pre-amplifier and the digital voltmeter. Power supplies for the amplifier may either be purchased as standard commercial items or the reader may prefer to construct a simple design such as that in Fig. 6.4 for himself.

References

1 Owens, A.R. (1982), *J. Phys. E.*, **53**, 789.
2 Savitzky, A. and Golay, M.J.E. (1964), *Anal. Chem.*, **36**, 1627.
3 Hicks, P.J., Daniel, S., Wallbank, B. and Comer, J. (1980), *J. Phys. E.*, **13**, 713.
4 Torr, M.R. and Devlin, J. (1982), *Appl. Optics*, **21**, 3091.

General reading

For further information, readers are recommended to look at the numerous handbooks and application notes published by the manufacturers of the various devices. Companies that can be consulted about operational amplifiers include Analog Devices, Burr-Brown, Motorola, National Semiconductor, and Texas Instruments, while those specializing in detectors include E.G. & G. Reticon, Hamamatsu, Integrated Photomatrix, Mullard, RCA and Thorn–EMI.

7 Interfacing techniques

7.1   Introduction

The increasing availability of cheap computing power has made the connection of computers to spectrometers and other laboratory instrumentation more and more desirable. The advantages to be gained range from the removal of error-prone manual measurement of spectral profiles and their entry into a computer via a punch card or keyboard, to the application of powerful computational enhancement of spectra at the spectrometer.

Unfortunately, computers and spectrometers generate and use data in fundamentally different ways, and in order for them to communicate it is necessary for an *interface* to be inserted between them. There are two basic tasks of an interface, no matter what application it is put to:

(a) It must ensure that the data generated by one instrument is converted correctly into a form that can be accepted by another.

(b) It must carry information between the two devices that shows each what the other is doing.

At the simplest, the transmission of data can be achieved with a single piece of wire, if it is not necessary to adjust voltage levels or signal type between the two devices. A more complex interface would change the voltage levels between the two devices and an even more complex interface would convert an analogue signal into a digital form. In the case of a spectrometer and a computer, the computer will require its data to be presented in a parallel, digital form (the binary bits of information representing the data are presented to the computer simultaneously, usually in 8-bit bytes on eight pieces of wire) while the spectrometer will produce its data as a simple analogue voltage.

It is also crucial to the correct functioning of the two devices that each knows what the other is doing, and provision of this information is another important function of the interface. There is no point in a spectrometer trying send a piece of information to a computer if the latter is busy at another task, and given the computational speed of modern computers, there is no point in a computer being idle and waiting for a slow spectrometer if there are other calculations it can be carrying out. The detail of how these two functions are implemented forms the basis of interfacing techniques.

It will have been gathered from the foregoing that there could be a multitude of different interfaces – one for each application. Indeed, up to about ten years ago this was probably true. Since then, the introduction of the microcomputer, and the wish to connect it to a wide range of devices, has brought about the introduction of a small set of interface types, some of which are the subject of international agreement on their definition. These interface types are primarily concerned with the transmission of digital information and consequently instrument manufacturers are increasingly incorporating into their instruments the necessary circuitry to convert analogue and non-standard digital signals into a form compatible with one of the defined standards. As most microcomputers also use at least one of the standard interface systems, connection of a computer to a spectrometer may only involve a simple cable connection.

Before discussing the detail of some of the interface types it is worth considering some general points about interfacing. In principle the interface sits between the computer and instrument (Fig. 7.1a) and all signals and lines needing conversion are fed to it. In practice it is much easier to split the interface into two parts: one that does a conversion of the computer signal into a defined state, and one that converts the instrument signal into the same defined state. Each piece is then inserted into the computer or instrument and connected by a cable carrying the defined signals (Fig. 7.1b). The advantage of this approach is that it makes the connection of different types of instrument and computers a simple task – to the outside world each talks the same language. The interconnecting cable is then described as the *interface bus*.

In general the interface bus will be a group of wires and these can be divided into three main classes: data information, address information and control information. Frequently the data and address information is multiplexed into one group of wires with the state of one control line indicating whether data or address information is on the

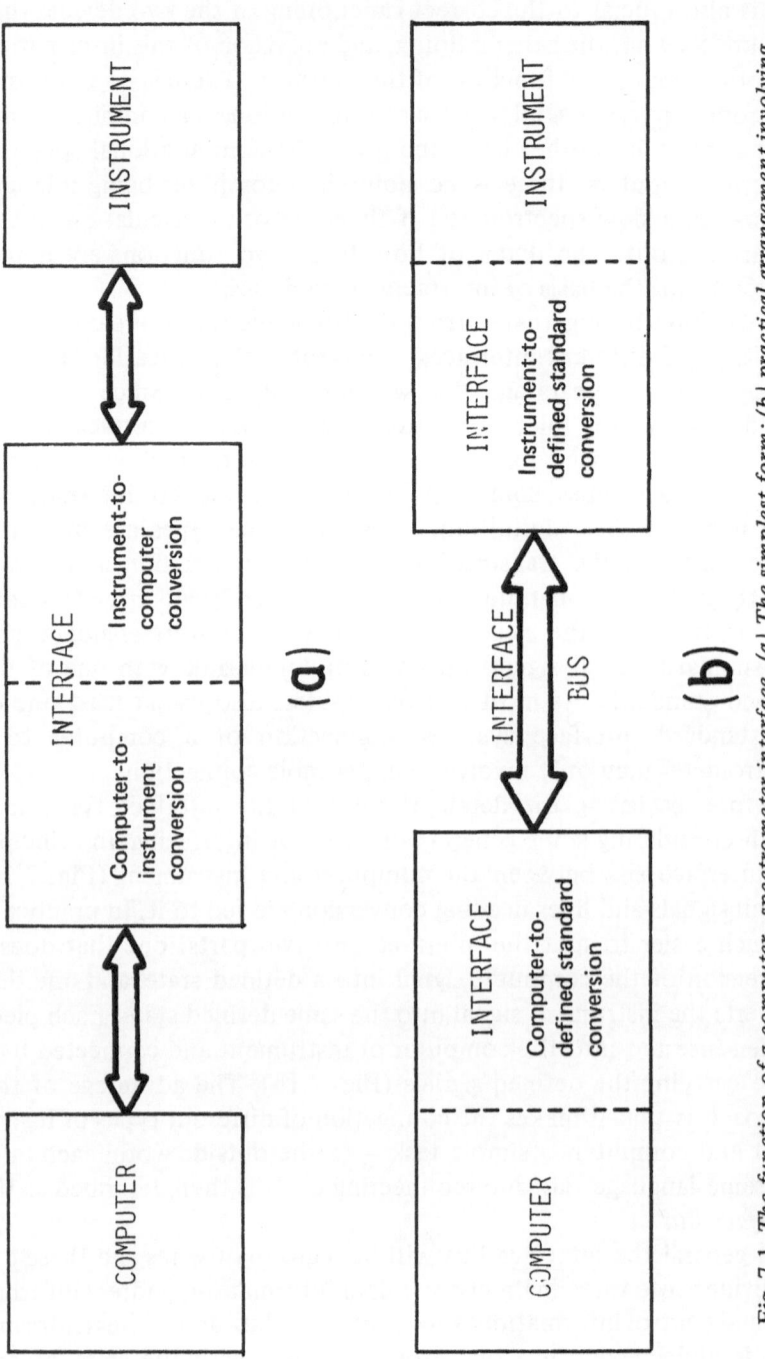

Fig. 7.1 *The function of a computer–spectrometer interface.* (a) *The simplest form;* (b) *practical arrangement involving connection by a common bus.*

lines. An important function of the control lines is to indicate to the devices on the bus the status of activity of each device. Fig. 7.2 illustrates a simple case of two devices with two control lines and device A wanting to read data from device B. For this example, a logic high on either line means that the message assigned to the line is active. The sequence of events starts with both lines low and then device A sets the 'I am ready for data' line high. This is detected by device B which responds by placing data on the data lines and a short while later sets the 'I have sent data' line high. On detecting this, device A sets 'I am ready for data' line low and reads the data from the data line. Setting this control line low forces device B to reset the 'I have sent data' line to low. The sequence can now be repeated for more data. The whole sequence using the two control lines is called *handshaking* and the data transfer is *asynchronous* as neither device is relying on the synchronization of the events of transmission and receipt of data. Some very high-speed interface systems must use *synchronous* transmission of data and thus have at least one control line carrying synchronization pulses.

Three types of interface will be described: the IEEE488, the serial interface and the BCD. The first two are standard interfaces, while the third is still a common form of data presentation in instruments.

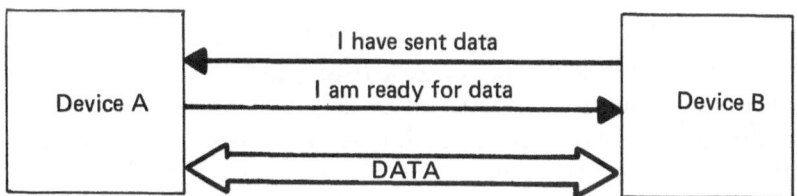

Fig. 7.2 *An illustration of handshaking between two devices, which could be computer and spectrometer.*

7.2   The IEEE488 interface

7.2.1   *General*

The designation IEEE488 refers to a standard set by the Institute of Electrical and Electronic Engineers [1] and is an 8-bit parallel interface with an additional eight lines communicating control information. The standard specifies a particular form of 24-pin connector and the appropriate pin assignments within the connector (Table 7.1) as well as the electrical and timing characteristics of the interfaces.

Table 7.1: *Connector and line designations for the IEEE488 interface bus.*

Connector pin no.	Function mnemonic	Function
1	DIO1	Data bit 1
2	DIO2	Data bit 2
3	DIO3	Data bit 3
4	DIO4	Data bit 4
5	EOI	End or Identify
6	DAV	Data Valid
7	NRFD	Not Ready for Data
8	NDAC	Not Data Accepted
9	IFC	Interface Clear
10	SRQ	Service Request
11	ATN	Attention
12	SHIELD	
13	DIO5	Data bit 5
14	DIO6	Data bit 6
15	DIO7	Data bit 7
16	DIO8	Data bit 8
17	REN	Remote Enable
18–24	–	Function not defined – may be connected to ground

The IEEE488 standard is used extensively by Hewlett Packard who first devised the system, and it is referred to by them as **HP-IB**. Many other manufacturers are now fitting the IEEE488 interface to their instruments. It is a particularly easy and convenient system to use providing the computer acting as controller has the necessary software built into the high-level language being used.

Because it is a parallel bus system the IEEE488 interface is used for the short distance (up to 20 m) communication between several devices (maximum 15) at quite high speeds. It is an *asynchronous system*, that is, after one byte of data been put onto the bus the next byte is not put on until all receiving devices have indicated that they have received the first byte and are ready for the next. Thus the maximum transmission rate along the bus is determined by the slowest active device. This may be a slow printer but could be the processor if it is communicating with a very high speed device. At very high data rates (above 100 kHz), cable impedance causing signal degradation is the limiting factor. Each device on the bus must have a unique address in the range 0 to 31. As this must be hardwired into the interface, it is not convenient to change it should there be a clash of addresses. Some manufacturers ease this problem by mounting a set switches on the instrument and the on-off pattern of these determines the address.

In the IEEE488 interface system, one device and only one device must be capable of assuming control of the bus at any stage. The device is termed the *system controller* and only it is allowed to reset all devices connected to the interface bus by setting the Interface Clear (IFC) line in the bus. The system controller may set the IFC at any time, even if it has passed active control of the bus to another device, and all devices connected to the bus will respond immediately by breaking off from any current tasks. Because of the importance of the system controller, the ability to set the IFC is hardwired into its circuitry and is absent in all non-system controllers. The system controller is normally the *active controller* of the bus system, although it may pass active control to any suitable device on the bus. Such an active controller takes control of all the bus management apart from the ability to set the IFC. Communications between the devices on the bus can be quite complex, but in all cases the basic sequence of events is the same.

7.2.2   *Active controller – device communication*

The active controller tests the 'Not Ready for Data' line (NRFD). If it is low then at least one device on the bus is not in a state to listen to the active controller. There is thus no point in the active controller sending a message. Once NRFD has returned to a high state, the active controller can place the first byte of a message on to the data lines DIO1–DIO8. After a short time to allow the data lines to settle, the active controller sets the 'Data is Valid' (DAV) line low to indicate to devices that there is data on the lines that can be read. On receiving the DAV low, all listening devices set their 'Not Data Accepted' (NDAC) line low showing that they are in the process of accepting the data and during this time they will set NRFD low. As each device accepts the data it will release its NDAC to high and when all devices have accepted the data the NDAC at the active controller becomes high. After accepting the data, each device may have to process it further before accepting another byte and it is only after this that the device will release NRFD to high. The active controller will not place another byte on the line while either NRFD or NDAC is low. This sequence of events, the handshaking sequence, will continue until the complete message has been transmitted.

The message transmitted can be one of two types: data or control information. The former can be, for example, an instrument reading or a line of text for printing. The latter is control information for the device and must be interpreted by the device as such, for example it can be a command for the device to listen to the data about to be transmitted or for the device to switch measurement ranges. The distinction between the two types of message is signalled by the 'Attention' (ATN) line of the bus. If ATN is high the information on DI01—DI08 is data, if it is low the information is to be interpreted as control or device address. The ATN line is controlled by the active controller. In the control/address mode the byte of information on the data lines is interpreted according to its value (Table 7.2). This means that device listen and talk addresses can be unambiguously and freely mixed with bus and device dependent commands; it also means that the active controller can address itself on the bus to listen or talk, and indeed it is essential that it does so for correct operation. An example of bus messages required to transmit data from an active controller to a printer is given in Table 7.3.

Much more complex messages may be required in some applications, but it is not always necessary for the user to program such

Table 7.2: *IEEE488 data value groups.*

Decimal value of byte	Meaning
0—31	Interface control commands
1	Goto Local (GTL)
4	Selected Device Clear (SDC)
5	Parallel Poll Configure (PPC)
8	Group Execute Trigger (GET)
9	Take Control (TCT)
17	Local Lockout (LLO)
20	Device Clear (DCL)
21	Parallel Poll Unconfigure (PPU)
24	Serial Poll Enable (SPE)
25	Serial Poll Disable (SPD)
32—62	Listen Addresses
63	Unlisten all devices
64—94	Talk Addresses
95	Untalk all devices
96—126	Device Commands (these are device-specific)
127	Delete

Table 7.3: *Example of IEEE488 interface level command sequence.*

Send from the computer the string 'HELLO' to a printer

Decimal value	ASCII equivalent	Interface command
		Set ATN low
85	U	Computer talk address
63	?	Unlisten all devices
42	*	Printer listen address
		Set ATN high
72	H	
69	E	
76	L	
76	L	Transmitted message
79	O	
13	Carriage return	
10	Line feed	

In a high level language supporting the IEEE488 interface this may appear as:

    PRINT # 1, "HELLO"
    or
    OUTPUT 705, "HELLO"

messages: high-level languages supporting the IEEE488 bus will generally interpret a simple high-level command into the required bus commands.

7.2.3 Interrupts

Communication between devices on the bus using the above handshake sequence is satisfactory for many applications, but there are others, however, when an interrupt-driven communication is appropriate, for example when the active controller has asked a device to carry out a sequence of measurements and report the result when it is available. During the time of the measurement the active controller can be doing other useful things.

The interrupt facility is provided by the 'Service Request' (SRQ) line. This line is normally held high but any device on the bus can pull it low when it wants the attention of the active controller. As more than one device can simultaneously hold the SRQ low, the active controller requires a mechanism of determining which devices are calling for attention. The bus message 'Serial Poll Enable' (SPE) achieves this; the active controller addresses each device on the bus in turn and asks it to send its serial poll status byte. This is read by the active controller and if bit 6 of the byte is set, the device has initiated a 'Service Request'. The action then taken by the active controller is completely device independent.

7.3   The serial interface

7.3.1   *General information*

There are three serial interface types in common use. They are desig-nated the RS232C(2), the V24(3) and the 20 mA current loop types. Historically, the 20 mA current loop system is the oldest and was developed to facilitate communications between slow electromechan-ical devices using changes of current direction to indicate high or low logic states. The RS232C and the V24 systems use changes of voltage level to indicate logic states and as such are more suited to higher data transfer rates than the 20 mA current loop system but over shorter distances. All three types of serial interface use the same basic logic of communication but there are differences in the detailed implementations. RS232C and V24 systems are usually interchange-able if the full communications sets are not used. The RS232C and V24 systems use nominally +12 V and −12 V to indicate logic low and high, respectively, although the standard for the circuits requires them to be designed to operate on very wide voltage ranges, for example +25 V to +3 V for logic low and −3 V to −25 V for logic high.

As with the parallel IEEE488 interface, the serial interface has two groups of signal lines, the data group and the control group (Table 7.4). Unlike the IEEE488 system, all signal lines in the serial interfaces are unidirectional, that is a device will send a signal out on one line and receive a response to it on another. Such a system has developed because of the requirement in long distance data communications to connect only two devices together and indeed it is still true that only two devices can be connected using a single serial interface. Again, unlike the IEEE488 system there are only two data lines, one for transmission and one for reception. This means that a byte of data must be transmitted in bit-serial form along one of the lines (Fig. 7.3). The conversion from a parallel byte of data within a computer to bit-serial data on the interface lines takes place within the interfaces at the transmitter and receiver and is not user visible.

Table 7.4: *Line assignment for serial interfaces.*

Pin no.	Mnemonic	Function
1	Ov*	Protective ground − a.c. power/equipment frame
2	Tx*	Data originated by the DTE for transmission for the DCE
3	Rx*	Data received by the DTE from the DCE
4	RTS*	'Request to Send' − tells the DCE that the DTE is ready to send data
5	CTS*	'Clear to Send' − tells the DTE that the DCE can accept data
6	DSR*	Tells the DTE that the DCE is not in a test mode and that its power is ON
7	Ov*	Signal ground − the signal reference level that must be common to the DTE and the DCE
8	RLSD	Tells the DTE that the DCE is receiving carrier signals
9	−	Reserved for test
10	−	Reserved for test
11	−	Unassigned
12	SRLD	Tells the DTE that the DCE is receiving secondary carrier signals
13	SCS	Tells the DTE that the DCE is ready to transmit data on the secondary channel
14	STD	Data from the DTE to be transmitted by the secondary channel of the DCE
15	TSET	Signal from the DCE to the DTE providing signal element timing information
16	SRD	Data from the secondary channel of the DCE
17	RSET	Signal to the receiving terminal providing signal element timing information
18	−	Unassigned
19	SRS	Tells the DCE that the sending terminal is ready to transmit data
20	DTR*	Tells the DCE that the DTE is ready to receive and transmit data
21	SQD	Signal from the DCE telling whether a defined error rate has been exceeded
22	RI	Signal from the DCE that a ringing signal is being received
23	DSRS	Selects one of two signalling rates in DCEs having two rates
24	TET	Transit clock signal provided by the DTE

* These lines are the ones commonly implemented in instrument RS232C interfaces.

Although the RS232C and V24 standards specify the electrical characteristics of the two systems and the designations of the lines used in the full implementation of the systems, it is rare to find two serial communication devices that can be connected together and be able to communicate immediately. There are a number of reasons for

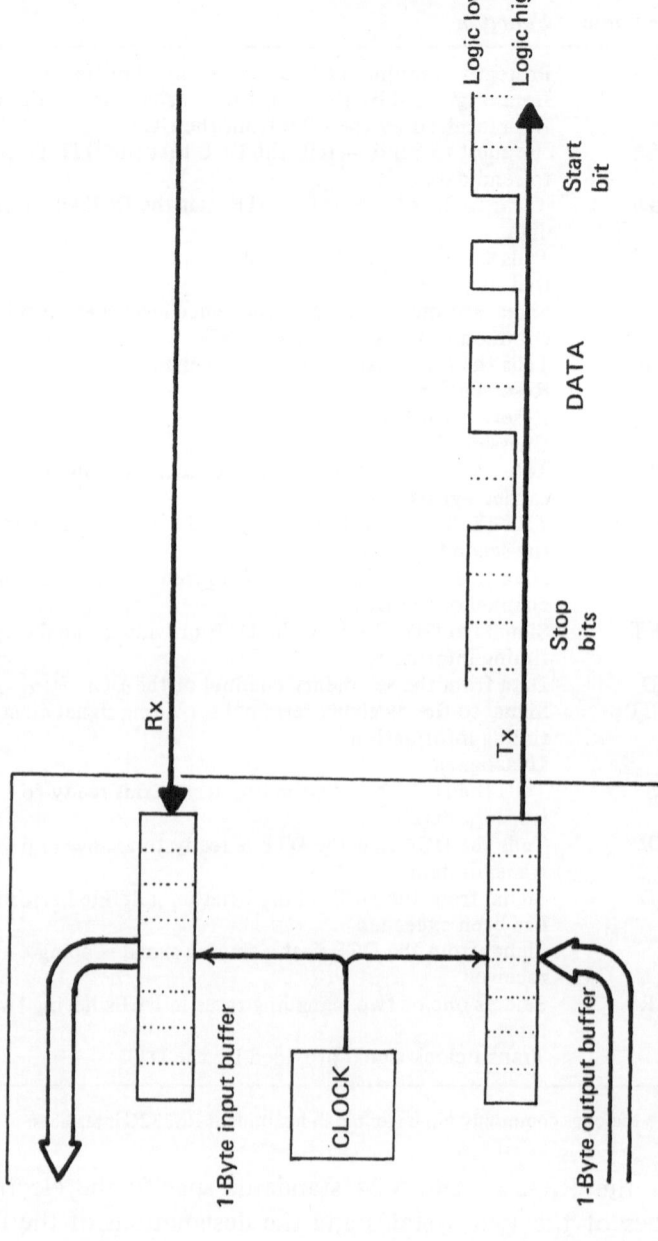

Fig. 7.3 The arrangement of a serial interface which used only two data lines. The data is transmitted in bit-serial form along one of the lines.

this but the basic one is that both standards are designed for data communications over long distances using modem equipment; in the laboratory the intermediate modem is not required. It is vital that anyone planning to set up serial communications between an instrument and a computer understands the problems: the fact that two devices are labelled RS232C does not mean they will communicate with each other.

7.3.2 *Transmitter and receiver*

Because of the unidirectional nature of the lines in the serial bus it is not easy to define one device as the transmitter and the other as the receiver, corresponding to active controller and non-active controller in the IEEE488 case. Following from the historical use of the serial system, a device is considered to be an 'active controller' if information enters or leaves the system through it and it is called the 'Data Terminal Equipment' (DTE). If the device can be considered to be communicating information within the system it is called the 'Data Communications Equipment' (DCE). The DTE generally has a 25-pin EIA plug on its cable with the transmitted data (Tx) on pin 2 (its corresponding received data (Rx) is on pin 3). The DCE generally has a 25-pin EIA socket on its cable with its received data (Rx) on pin 2 and its transmitted data (Tx) on pin 3. Such devices *can* be connected together, but beyond the difficulty of defining DTE and DCE for laboratory instruments, many manufacturers do not follow the connector convention and it is possible to find that a 25-pin EIA plug has been wired as DCE and a 25-pin EIA socket has been wired as DTE. Such problems apply also to the pin designations of the handshake lines.

7.3.3 *Handshake lines*

The RS232C and V24 specifications are devised for long distance data communications. For this it has been necessary to designate a large number of lines to indicate the status of one device to the other and the status of the communications link to both devices. In most laboratory applications this plethora of information is not required so most instrument manufacturers implement only a subset of the full serial communications specifications. Unfortunately, the subsets are not standardized and vary from a simple three-wire system (Rx, Tx, and signal ground) to a ten-wire system. More commonly a five-wire system is used: Rx, Tx, ground and two handshake lines. But as the full system can be considered to use two pairs of handshake lines there is no guarantee that the handshake lines in two five-wire inter-faces are compatible, even if the correct DTE and DCE connectors and pin designations are used.

The two pairs of handshake lines are the 'Request to Send' (RTS)/ 'Clear to Send' (CTS) pair on pins 4 and 5 and the 'Data Set Ready' (DSR)/'Data Terminal Ready' (DTR) pair on pins 6 and 20. The DSR is strictly not a handshake line but it is used by some suppliers as such.

In addition to these problems it is not uncommon to find that a manufacturer has internally connected the handshake lines in his interface to guarantee continuous transmission or reception of data. Three-wire serial systems are almost invariably of this type.

7.3.4   *Data transmission rate*

As in the IEEE488 system, bytes of data in the serial system are communicated in an asynchronous fashion using the handshake lines. However, because there is only one wire along which to send the byte of information, each bit forming the byte must be placed on the wire one after the other. This is the so-called bit-serial method of data transmission (abbreviated to serial transmission). Conversion of the byte into bit-serial, and the reverse operation, is a function of the interface and it is achieved by using an internal clock to step a byte on to the line one bit at a time at a specified rate. The rate is known as the baud rate and this is simply the bit rate in bits per second. It is not associated with the rate at which bytes are transmitted as this is generally done asynchronously.

For two devices to communicate, their internal clocks must be adjusted to operate at the same baud rate. About a dozen different baud rates are in common use ranging from 110 to 9600 baud but not all devices are capable of covering this range and the probability is that any two devices chosen to communicate with each other will be set at different rates. Baud rate selection is commonly achieved by switch settings on the interface card, but many systems now offer a selection under program control.

7.3.5   *Start, stop and parity bits*

To allow the receiving interface to recognize a byte of information being received and to synchronize the 'ticks' of its clock with those of the transmitting clock, the transmitting interface adds to the byte one bit of logic low at the start of the byte and either one or two bits of logic low at the end of the byte. These are the 'start' and 'stop' bits. There will always be one 'start' bit, but it is clearly necessary to know if one or two 'stop' bits have been used when bytes of data follow each other closely.

The parity bit is the most significant bit of the byte transmitted. Parity may or may not be used in serial communications, but if it is, it is used to check that the byte of information has been transmitted and received correctly. If parity is used, then the information can only occupy seven bits of the byte; if parity is not used the information can occupy the eight bits of the byte. The choices available are:

Even parity: The seven one-bit values of the byte are added, and if the least significant bit of the result is zero, the parity bit is set.

Odd parity: The seven one-bit values of the byte are added, and if the least significant bit of the result is one, the parity bit is set.

Zero parity: The parity bit is always set to zero.

One parity: The parity bit is always set to one.

No parity: The most significant bit of the byte does not indicate parity and is part of the information.

The two interfaces communicating with each other must be set to have corresponding stop and parity bit configurations, and again this is done either by switch settings or under program control.

7.3.6 *Software handshakes*

The communications protocols used with serial transmission of data allow for the use of software handshakes instead of the hardware handshakes described in Section 7.3.3. Software handshakes are achieved by reserving two characters to indicate 'I am ready to receive data' and 'Data transmission is complete'. The first character is transmitted as an ordinary byte by a device waiting to receive data, the last is added to the data stream as the last byte. Devices that recognize these bytes can then operate a handshaking procedure for the communication of data.

7.3.7 *Interrupts*

The ability of one device in a serial link to interrupt the operation of the other is not a defined part of the system. However, many serial interfaces are built so that the operation of the handshake lines or the receipt or transmission of a byte of data can generate an interrupt.

The use of interrupts with serial interfaces is totally dependent on the computer and interface design, and reference must be made to appropriate manuals for details.

7.4   The BCD interface

The Binary Coded Decimal (BCD) interface is completely non-standard, although BCD outputs abound in instrumentation. The name arises from the representation of each decimal digit of a number by its binary equivalent in four bits. Thus the number 936 is represented in BCD as 1001 0011 0110. Four bits for each digit must be used as the leading zeros are significant in decoding the number. A BCD connector on an instrument will certainly provide in parallel the binary representations of all significant decimal digits that are to be transmitted. Information relating to the sign, decimal point position, overrange and underrange will almost certainly be provided either as single bits or as a small group of bits. Two control bits may be provided corresponding to the 'Data is Valid' (DAV) and 'Not Ready for Data' (NRFD) lines of the IEEE488 interface or the RTS and CTS lines of the serial interface. All of this information will be available, in parallel, at a single connector of an undefined type. The number of bits to be accommodated can be between about 8 and about 64.

Some computer systems have general purpose BCD interfaces available to ease the communication of BCD data from an instrument. Such interfaces must be configured to meet the requirements of the particular instrument. This may be done by moving hardware links, but preferably it will be done under software control from the computer. If a suitable general purpose BCD interface is not available it is possible to construct one to meet specific requirements. In this case the collection of the BCD data from the instrument is generally straightforward as there are many suitable integrated circuits available. However, transforming this data into a form acceptable to the computer and meeting all of the timing, addressing, and interrupt control requirements of the computer requires a detailed and thorough understanding of the computer and the proposed interface. All of the common microprocessor types, Intel 8080, Motorola 6800 and Zilog Z80 series for example, have compatible integrated circuits suitable for this work. In addition to the hardware problems, high-level languages generally do not support BCD data transfers and suitable low-level language device drivers may have to be written. Should the direct conversion of data by this method not be practicable, the simpler task of converting either the analogue or the BCD information into one of the standard interface types, IEEE488 or RS232C for example, can be considered. Again, single integrated circuits, or single board systems, are available at moderate cost to carry out these conversions.

7.5 Conclusions

Connecting an instrument to a computer requires a suitable interface to be inserted between the two. The ease with which this can be done varies considerably with the interface type used. If the instrument has one of the two standard interfaces, the IEEE488 or the RS232C/V24, connection is usually straightforward although the number of variables associated with the serial interfaces generally means that several attempts may have to be made in order to get the system to work. If the instrument has a non-standard interface, such as BCD, the connection can range from the trivial to a major electronic design and construction project.

The moves of instrument manufacturers towards providing one or both standard interface types is to be applauded and encouraged, for these days it is rare for a laboratory worker not to want to process further data emerging from an instrument.

References

1 Institute of Electrical and Electronic Engineers (IEEE), USA (1978), Standard 488.
2 Electronic Industries Association (EIA), USA (1969), Standard RS232C, 2001 Eye Street NW, Washington DC 20006.
3 Comité Consultatif Internationale Téléphonique et Télégraphique (CCITT), USA, (1969), Recommendation V.24, United Nation Bookstore, United Nations Assembly Building, New York NY 10017.

8 Cells and cell holders

8.1   Types of cell

The absorption spectra of a wide variety of sample types have been measured ranging from molten salts at high temperatures to labile free radicals trapped at liquid helium temperature. A review of these techniques is beyond the scope of this book, which is primarily concerned with the measurement of solutions at near-ambient temperature. Solutions must be put into a transparent *cell* or cuvette with a specified optical pathlength. Nearly all spectrometers are designed to accommodate a *normal* 10 mm cell (see Fig. 8.1) and, in the majority of applications, the concentration of the test solution can be adjusted to bring its absorbance in a 10 mm layer into an acceptable range.

When only a limited amount of solution is available, a *semi-micro* or *micro* cell can be used. Concentrated solutions that cannot be diluted may be measured in a short-pathlength cell, while on the other hand, gases or dilute solutions that cannot be concentrated require long-pathlength cells.

Some kinds of solid sample can be measured fairly easily: a large crystal or a polymer block or film that is free from optical defect can be measured if it has two parallel faces of reasonable optical quality. Such a sample will not require a cell but only some means of holding it firmly and perpendicular to the measuring beam. Alternatively, solid samples can be ground up and suspended in a medium of suitable viscosity and refractive index; they can then be treated like a solution. These techniques will be reviewed in Section 9.8. Reflectance spectra can also be measured from solid samples.

8.2   Choice of a cell

8.2.1   *The normal 10 mm cell*

Open-topped rectangular cells are the first choice for any solution measurement since they fit most instruments, are relatively cheap and are available in a variety of grades and materials of construction. Their use is subject to the following restrictions:

(a) The absorbance of the solution in a 10 mm layer is within the optimum range of the instrument, say 0.1 to 1.5 A. This will be discussed in Section 9.2 and more fully in Section 9.7.3.

(b) The solvent should not be volatile. Evaporation will increase the absorbance of the solution during measurement and so for most organic solvents, a cell with a lid or, better, a stopper should be used.

(c) A sufficient volume should be available. For most instruments, at least 2 ml is required in this type of cell for satisfactory measurements.

(d) There are only a few samples to be measured. For repetitive measurements, a 'sipper' or flow system using special cells is preferable.

Having chosen this type of cell, a decision must be made about the grade of cell and the material of construction suitable for a particular measurement. The cheapest cells are made of plastics which absorb strongly below 280–300 nm (see Appendix A2.2). Their low cost means that they are disposable and are thus ideal for samples that are dangerous or difficult to remove. Their optical quality may be adequate for routine measurements but not high precision work, particularly since their pathlength varies from top-to-bottom of the cell. They cannot, of course, be used with the majority of organic solvents.

Next in the price range come the glass cells: these can be of high optical quality, but even the best optical glass does not transmit below 300 nm. Fused natural quartz cells transmit much further into the UV region, but impurities limit their performance in the far-UV (see Appendix A2.1). The best cells are those of fused synthetic silica, which have effectively no absorption down to 200 nm, are robust and are resistant to attack by most chemicals.

Most manufacturers supply synthetic silica cells made to very good dimensional tolerances which are satisfactory for the majority of applications. Special cells with accurately determined pathlengths are available, and the National Bureau of Standards supplies cells with certified pathlengths. Recommended tolerances and the selection of cells are fully discussed in Volume 1 of this series [1].

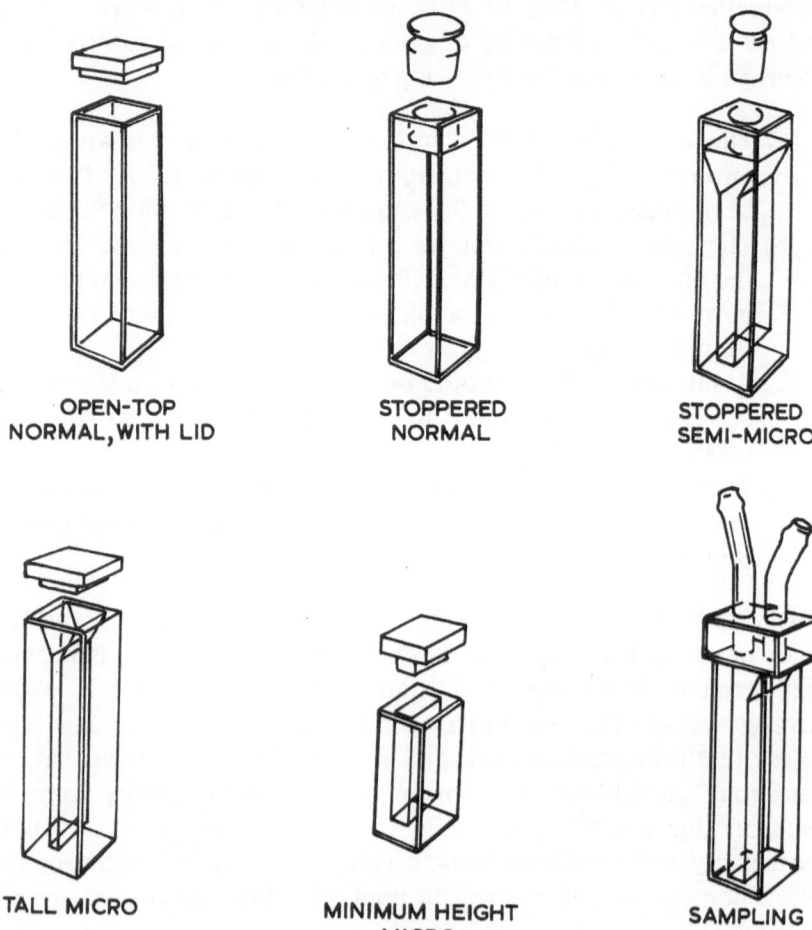

OPEN-TOP   STOPPERED   STOPPERED
NORMAL, WITH LID   NORMAL   SEMI-MICRO

TALL MICRO   MINIMUM HEIGHT   SAMPLING
                   MICRO

Fig. 8.1 *Representative types of spectrophotometer cell. Reproduced from [1].*

8.2.2   *Micro and semi-micro cells*

If only 0.5–2 ml solution is available, a semi-micro cell can be used. This has the same outside dimensions and the same pathlength as a normal 10 mm cell but the chamber is only about 4 mm wide. This means that the *working area* – the region of the cell window that the beam passes through – is smaller than that for the normal cell (Fig. 8.2). The measuring beam of most instruments is sufficiently narrow to be within the working area of a semi-micro cell, but users should apply the tests described in Section 9.4d when first using this type of cell in their instrument.

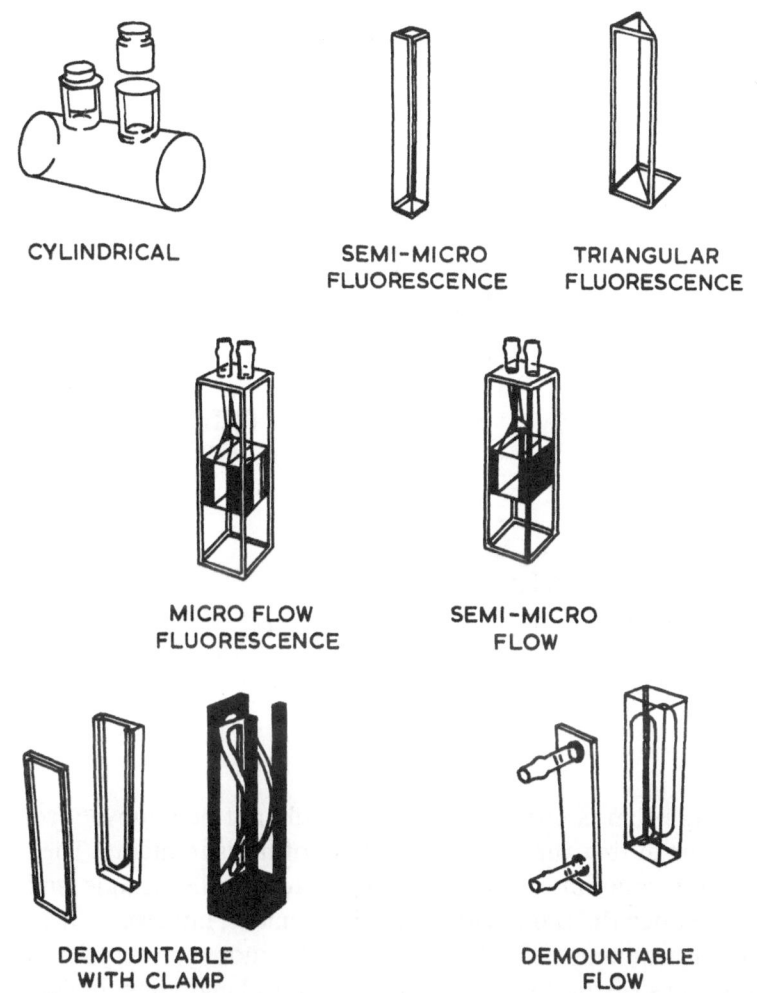

CYLINDRICAL

SEMI-MICRO
FLUORESCENCE

TRIANGULAR
FLUORESCENCE

MICRO FLOW
FLUORESCENCE

SEMI-MICRO
FLOW

DEMOUNTABLE
WITH CLAMP

DEMOUNTABLE
FLOW

A micro cell is necessary for volumes down to about 200 μl. The working area of these cells is less than 2 mm wide and so there is a real risk of part of the beam passing through the cell wall rather than through the sample. To overcome this, a special cell holder that is fitted with a mask slightly narrower than the working area should be used. If such a mask is not available, a mask of black card or metal foil can be fixed to the normal cell holder. Alternatively, 'self-masking' cells are available which have side walls made of black glass or silica which absorbs light passing outside the working area (Fig. 8.1). Whichever arrangement is used, the positioning of the cells in the

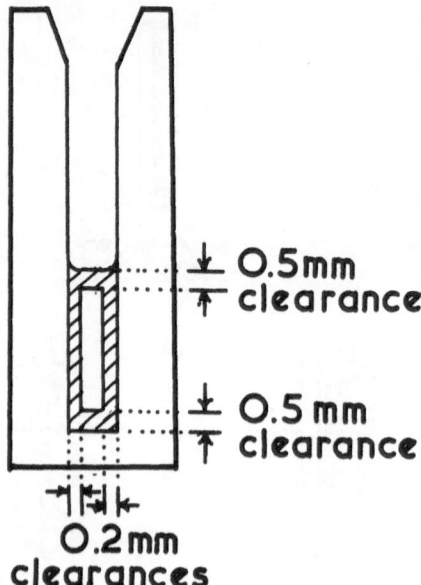

Fig. 8.2 *Diagram illustrating the working area of a cell: the minimum area of the windows over which the dimensions and optical properties of the cell are guaranteed by the manufacturer, and through which the measuring beam should pass. The minimum filling level is also shown.*

measuring beam is critical and they should not be removed from the instrument during the course of a series of measurements. The masks reduce the amount of light passing through the sample and thus slightly reduce the sensitivity of the instrument; however, their effect on the instrumental baseline is potentially more serious, for cutting off the sides of the beam may produce wavelength-dependent deviations.

8.2.3   *Cylindrical cells*

The measuring beam in most spectrometers converges or diverges to some extent as it passes through the cell housing. This means that for measurements requiring a long pathlength, say 40 mm or more, the working area of the cell must be larger than that of a normal rectangular cell. A cylindrical cell has a much larger working area and is preferable for such measurements (Fig. 8.1). Not all modern instruments can accommodate long cells, and few are supplied with holders for cylindrical cells.

8.2.4 *Tubular cells*

Some colorimeters and simple spectrometers use test-tube-shaped cells of 10 mm nominal internal diameter. Although they are cheap and robust, the uncertain pathlength and optical behaviour of these cells make them unsuitable for precise measurements.

8.2.5 *Flow cells*

There is a wide variety of flow cell types but they can be grouped into two classes:

(a) *Emptying cells*
Instruments with sipper devices for automatically filling and emptying cells require a cell that can be completely emptied by suction.

(b) *Continuous flow cells*
These are intended to monitor a stream of liquid without interruption. Each element of the stream must completely displace the preceding element, and so there must be no regions of low flow rate. In addition, there should be no traps for bubbles carried in by the stream. The volume of the cell must be as small as possible for a given pathlength — an extreme example is the HPLC detector cell, where the chamber is typically 10 mm long but only 1 mm in diameter, thus giving a volume of about 8 μl. Successful use of such a cell requires precise alignment of the measuring beam and care must be taken to avoid the effects of changes in the refractive index of the sample stream. The application of these cells will be described in Chapter 12.

The connection of tubing to the cell can be troublesome, particularly if the cell is to be changed frequently. A new system for the connection of tubing to glass or silica cells has recently been introduced, and is illustrated in Fig. 8.3.

8.2.6 *Demountable cells*

Normal cells with pathlengths down to about 1 mm are available for the measurement of high absorbance liquids, but cells as small as this are difficult to fill and even more difficult to empty and clean. The front window of a demountable cell can be readily removed, and a generally available pattern that will fit most cell holders is shown in Fig. 8.1. Another useful design is shown in Fig. 8.4 and is based on a standard IR cell. Two flat fused silica windows are separated by a

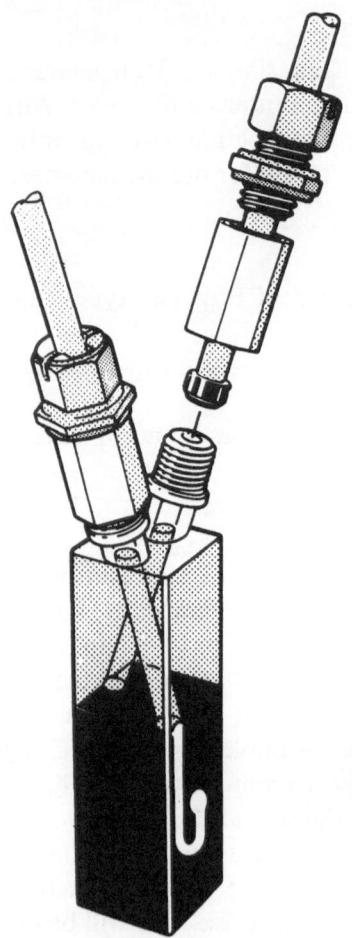

Fig. 8.3 *A novel means of connecting tubing to glass and fused silica flow cells. By courtesy of Starna Ltd.*

PTFE washer; washers of various thicknesses are available and so the pathlength can be adjusted to suit the sample. The cell is easily dismantled to clean the windows, and if these become contaminated, they can be replaced at a reasonable cost. Few instruments are provided with a holder for this type of cell, and so a holder must be made for your instrument.

8.3   Cell holders and sample compartments

Cells for special purposes can be purchased fairly cheaply, but the cell holder and the sample compartment are often an integral part of

Fig. 8.4 *A demountable variable-pathlength cell based on an IR cell. Another version of this cell has Luer ports so that it can be filled and emptied with an hypodermic syringe. By courtesy of Beckmann–RIIC Ltd.*

the spectrometer and it therefore is essential to anticipate the future uses of the instrument when deciding which one to buy. If the only task for the instrument is the measurement of solutions whose concentrations can be adjusted and where 2 ml solution is always available, then a spectrometer with a cell holder that will only take 10 mm normal cells will be satisfactory. However, most laboratories have to measure more demanding samples from time to time, and the following aspects of sample measurement should be borne in mind.

8.3.1   Temperature control

The absorbance of most solutions changes with temperature as the result of a number of factors including: (a) expansion of solvent, which is particularly troublesome with some organic solvents (see Appendix A1); (b) changes in the apparent molar absorptivity of the solute, which is often due to shifts in the equilibrium between different molecular forms of the solute; and (c) the decomposition or reaction of unstable solutes. As examples of these effects, the density of dioxan changes by 0.112% $^{\circ}C^{-1}$ at $20^{\circ}C$. The sample compartment of a spectrometer in a cold laboratory could rise in temperature by $10^{\circ}C$ as the instrument warms up, resulting in a 1.1% change in the measured absorbance of a dioxan solution. The molar absorptivity of the 350 nm peak of potassium dichromate in acid solution changes by -0.05% $^{\circ}C^{-1}$ over the range $20-30^{\circ}C$ [2]. A $10^{\circ}C$ rise in sample temperature could therefore result in a 0.5% drop in the absorbance of the solution. Therefore, all accurate measurements should be carried out at a defined temperature, and so some means of regulating the temperature of the cell and of measuring its temperature is desirable. Beyond the problems of absorbance changes, many assay techniques involve kinetic measurements which require very precise temperature control. The temperature of the cell compartment of all spectrometers rises to some extent above ambient while the instrument is running, and the modern trend of building-in the lamp housing to make a more compact instrument has not eased the problem.

Temperature control can be accomplished by thermostatting (a) the sample housing, (b) the cell holder, or (c) the cell itself. Of these, (b) and (c) give the best control, and (b) is the most common method. One snag with (b) is that thermostatted cell holders will often only accept 10 mm normal cells. A jacketted cell, such as that shown in Fig. 8.5 is a good alternative means of temperature regulation.

The precision of temperature control depends upon the type of measurement to be made. The problem of absorbance changes outlined at the beginning of this section require control to $\pm 0.5^{\circ}C$, but the determination of reaction rates can require control to $\pm 0.05^{\circ}C$.

Measurements at temperatures below $-10^{\circ}C$ require the use of a specially designed cryostat to house the sample. These are available from specialist instrument makers, who should be consulted before the spectrometer is chosen. Some special techniques based on the effects of temperature change — for example, the determination of the melting temperature of proteins — also require special accessories and it is prudent to choose these before selecting an instrument.

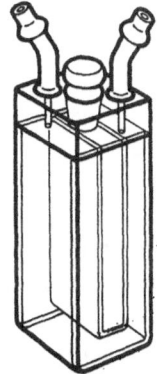

Fig. 8.5 *A silica cell of fused construction with an integral thermostatting jacket. The projecting tubes carry the thermostat water, and the cell is filled through the central stopper. By courtesy of Thermal Syndicate Ltd.*

8.3.2   *Micro-cell holders*

The satisfactory use of a micro-cell requires a cell holder with built-in masks and, if possible, some means of adjusting the cell laterally in the beam. The height of the beam above the floor of the cell varies between instruments, and this dimension becomes critical if only limited amounts of solutions are available. The beam should pass through the working area as shown in Fig. 8.2.

8.3.3   *Long-pathlength cells*

If gases or very dilute solutions are to be measured, seek out one of the spectrometers that will take cells up to 100 mm long.

8.3.4   *Sipper systems*

If a large number of samples are to be measured, a sipper system will speed the work, reduce damage to the instrument caused by spillages and will probably increase the accuracy of the measurements.

8.3.5   *Flow systems*

In general, flow cells are compact and will fit most sample compartments. It will, of course, be necessary to run tubing in and out of the compartment through light-tight seals. The tubing should also be opaque or it may act as a light guide. If an instrument is intended for a critical application such as HPLC monitoring, where a solution of low absorbance is measured in a cell of minute working area, it is important that the instrument is capable of passing a sufficient amount of energy through the cell to give a good signal-to-noise ratio, and that it is sufficiently stable and sensitive to produce satisfactory data. For such applications, arrange to test the instrument in the laboratory before purchase.

8.3.6   *Stirrers*

If a reaction is to be carried out in the cell, a mechanical stirrer can be of great benefit. A convenient method is by means of a small magnetic stirrer bar driven by a special coil placed in the cell holder beneath the cell (Fig. 8.6). The coil is sufficiently thin that the cell is not raised unduly, and is provided with a power supply that causes it to generate a rotating magnetic field. Before the device is used, checks should be made that the magnetic field does not upset the photomultiplier tubes in your instrument, and that the rotating stirrer bar does not interfere with the measuring beam. In the latter case, the bottom of the beam can be masked so that the stirrer bar cannot interrupt the beam.

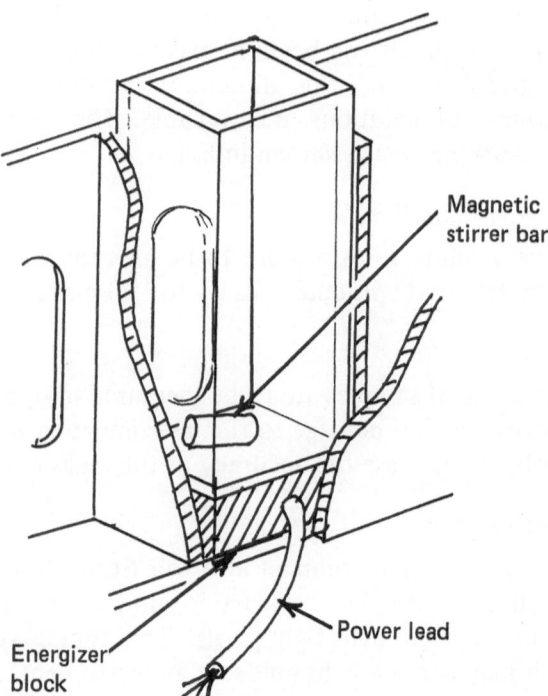

Fig. 8.6 *A miniature cell-stirring device. A rotating magnetic field generated in the block placed in the cell holder beneath the cell drives a small stirrer bar in the cell. A stirrer of this type is sold by Lawrence Instruments, Ontario, Canada.*

8.3.7 *Cell changers*

Kinetic measurements on a group of solutions can be simplified by the use of an automatic cell changer and timing unit, which can be programmed to measure the solutions in turn at specified wavelengths and time intervals.

8.3.8 *Fluorescence attachments*

Provision for the right-angle illumination of the sample with an excitation beam from a secondary light source is available with some instruments. Such accessories are unsatisfactory for a number of reasons, and a dedicated fluorimeter is always preferable.

8.3.9 *Reflectance accessories*

Integrating spheres and other attachments for the measurement of opaque samples are available for some instruments.

8.3.10 *Measurement of turbid or scattering samples*

Some instruments have provision for placing the cell near the window of the detector in order to minimize the effects of light-scattering by the sample. An additional aid is a built-in diffusing screen between sample and detector. The problems associated with such measurements will be discussed in Section 9.8.2.

8.3.11 *Gas purging*

Some instruments have provision for blowing out the monochromator and sample compartment with nitrogen or argon. This has two applications: (a) for measurements in the far-UV region where absorption of the measuring beam by oxygen reduces the sensitivity of the instrument, and (b) in measurements at low temperature, for if the cell is colder than about 10°C below ambient on a typical English day, condensation will form on the windows. This can be avoided by purging the sample compartment with dry gas.

8.3.12 *Cell location*

When appraising any instrument, the cell holder should be examined carefully, for this is central to the production of accurate and reproducible data. It should be placed so that the cell can be filled and emptied easily. The means by which the cell is located in the holder and the holder in the instrument are important: the cell should be located by the front face and one side held in firm contact with

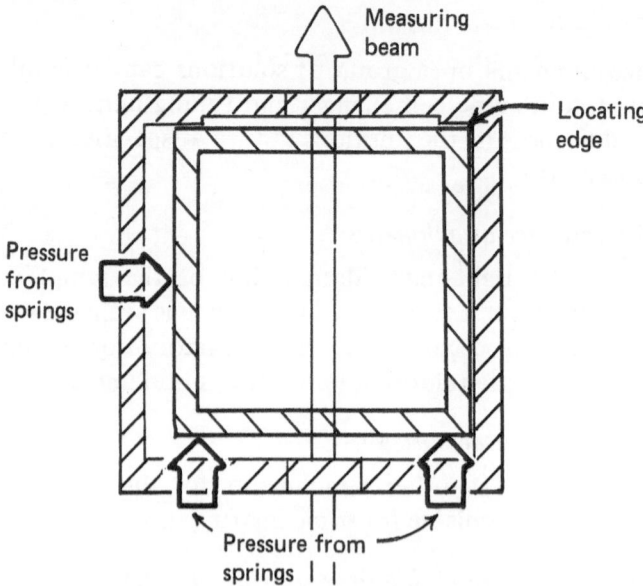

Fig. 8.7 *Plan view illustrating how a cell should be located in the cell holder.*

locating surfaces, although there should be no risk of scratching the working area when the cell is removed from the holder. Fig. 8.7 shows a suitable means of locating a rectangular cell. The springs should be strong enough to ensure that the cell is properly located, but not too strong as this will make it difficult to remove the cell.

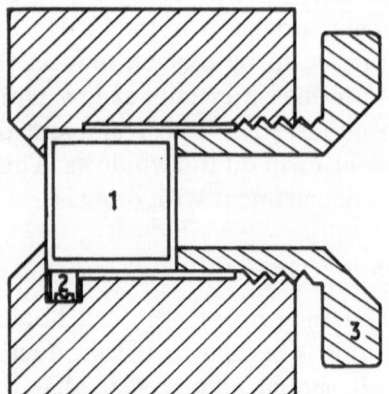

Fig. 8.8 *A special cell holder in use at the National Physical Laboratory. The cell (1) is held against the locating face by a spring (2) and clamped in position by a threaded cylinder (3). Reproduced from [1].*

The holder should not sit on a flat surface in the instrument but should be raised on some kind of self-locating mount, because otherwise dust or corrosion beneath the holder will cause it to rock.

For measurements of the greatest precision, it was recommended in Volume 1 of this series that a special holder should be made similar to that shown in Fig. 8.8. The cell is clamped in position by means of a screw, and is filled and emptied *in situ*, only being removed at the end of a series of measurements.

References

1 Burgess, C. and Knowles, A. (Eds) (1981), *Techniques in Visible and Ultra-violet Spectrometry*, Vol. 1, *Standards in Absorption Spectrometry*, Chapman and Hall, London.
2 Burke, R.W. and Mavrodineanau, R. (1977), *NBS Special Publication*, No. 260–54.

9 Measuring the spectrum

This chapter is primarily concerned with the measurement of the absorption spectra of clear solutions; techniques for dealing with more difficult samples will be briefly dealt with in Section 9.8.

9.1   Choice of solvent

When presented with a solid sample, a determined attempt should be made to get all of it into solution. The nature of the compound should be considered. For organic compounds, the rule of 'like dissolves like' should be remembered and so the polarity of the solvent should match that of the sample. Thus a hydrocarbon will probably only dissolve in a hydrocarbon solvent, while polar molcules require polar solvents. A solvent can be found for most low-molecular weight organic molecules. The properties of some useful solvents are given in Appendix A1. Higher molecular weight compounds present greater problems and some proteins and polymers will not form true solutions in any solvent. Many biological macromolecules will give clear micellar suspensions in detergents. If you are unsure about the choice of a suitable solvent, do not hesitate to refer to the originator of the sample — he is the person most likely to understand the problem.

Beyond the fundamental question of its ability to dissolve the samples, the following points should also be considered when choosing a solvent.

(a)   *Transmission*
The solvent must have a high transmittance in the spectral region of interest. This is particularly important when selecting polar organic solvents for use below 300 nm. The useful ranges of some common solvents are given in Appendix A1. Although distilled water transmits satisfactorily down below 200 nm, it should be remembered that detergents and buffer salts can absorb strongly at wavelengths up to

300 nm. Other less obvious transmission problems should also be borne in mind. Many organic solvents contain antioxidants or inhibitors that absorb at longer wavelengths than the solvent itself and so, if the latter is allowed to evaporate, the apparent cut-off wavelength can change. Oxygen and other gases dissolve in water and many organic solvents, and this will affect solvent transmission at shorter wavelengths; consequently routine purging of solutions with nitrogen or argon is advisable for measurements in the far-UV.

(b) *Handling problems*
Highly volatile solvents such as diethyl ether should be avoided at all costs, and alkanes smaller than hexane should not be used for accurate work. Even with a 'standard' solvent such as ethanol extreme care should be taken to avoid changes in concentration due to evaporation. Solvents such as benzene, ether, chloroform and carbon tetrachloride should also be avoided if possible as they are health hazards.

(c) *Interaction with the solute*
The possibility of chemical reaction or complex formation with the solvent should be considered. Polar organic solvents nearly always cause shifts of spectral maxima and in some cases these are very large. Similarly, changes in the pH of the solution can cause protonation or deprotonation of the solute giving rise to major changes in the absorption spectrum. When making comparisons with standard spectra it is essential that the solvent composition, pH, ionic concentration etc., are identical if valid conclusions are to be reached.

9.2 Making a solution

Having chosen a solvent, it is now necessary to decide upon a suitable concentration for the test solution. As a rule of thumb, for low-molecular weight organic molecules, 1 mg dissolved in 10 ml solvent should give a sufficiently concentrated solution for the measurement of the major maxima; for example, if the molecular weight of a compound is 300 and the molar absorptivity of a typical peak is 2×10^3 M^{-1} cm^{-1}, then the solution in a 10 mm cell will give a peak absorbance of

$$A = \epsilon bc = 2 \times 10^3 \times 1 \times \frac{1 \times 10^{-3}}{300 \times 0.01} = 0.67$$

which is a suitable value for measurement.

The optimum absorbance range for a particular instrument is determined by a number of factors which will be discussed in Section 9.7. The optimum range for most modern instruments is about 0.3−1.5 A. Older single-beam instruments using photocells rather than photomultipliers generally have a lower optimum range of 0.15−0.7 A. The critical parts of the measured spectrum should be kept within the appropriate limits.

The solute should be weighed out as accurately as possible. This may be considered unnecessary if the identification of the solute is the only purpose of running the spectrum; however, molar absorptivity values for the peaks may also be of help in identification or give some clues about the purity of the material. It is often necessary to repeat the determination at a later date and quantitative information then becomes valuable.

For the same reasons, the solute should be dissolved as completely as possible in an accurately defined volume of solution. These two aims cannot always be achieved in one operation! If you are confident that the compound will dissolve readily at room temperature, then the solution can be made up in a volumetric flask. The solid is put into the flask using a small glass funnel and the residue washed into the flask using about 80% of the final volume of solvent. The flask is stoppered, shaken until all the solute is dissolved and then made up to the mark. For unknown or less soluble material, the solution must be made before it is put into the volumetric flask. Use a beaker or conical flask, which can be stirred more conveniently and can be heated if necessary: Never heat volumetric glassware for it may never return to its original volume. For granular material, a good technique is to put the solid in a small agate mortar, add a small amount of solvent and grind the solid to a paste. Further amounts of solvent are added and mixed in. The undissolved material is allowed to settle, the supernatant transferred to a volumetric flask using a Pasteur pipette and the procedure repeated until the solid is all dissolved.

Dissolution can be speeded by heating the solvent but care should be taken that the solute is not affected by this. Often prolonged mechanical shaking or stirring at room temperature will succeed with apparently insoluble compounds. Immersion of the flask in an ultrasonic clearing bath is an excellent means of speeding dissolution. If the weighed sample fails to dissolve completely, decant off the solution and add further solvent to the residue. The possibility that the original material was not homogeneous should be considered, particularly if the second batch of solvent fails to dissolve any more of it.

The success of the solution is usually judged by eye, but it is easy to be deceived by colourless compounds of refractive index near to that of the solvent. If there is any doubt, measure the spectrum, filter or centrifuge the solution, and measure again to see whether any change has occurred. If so, particles were present in the solution and its concentration is less than supposed.

9.3   The cell

The choice of cell for a particular task was dealt with in Chapter 8 and is, of course, related to the choice of solution concentration. The normal open-topped 10 mm cell is always first choice, for it is easy to handle, to fill, empty and clean. They are, however, easily knocked over when standing on a bench and should always be stood in a small beaker or fitted into the cell holder before being filled. The packaging for plastic cells can form a useful stand. All double-beam and some single-beam instruments need a second 'reference' cell containing 'solvent' − that is, all the components of the sample solution except the sample itself. The purpose of this cell is to compensate for the absorption, reflection and refraction of the sample cell and the solvent, and so the optical characteristics of the cells should match as closely as possible. Opinions vary about the necessary precision of this match. Traditionally glass and natural quartz cells were sold in 'matched pairs' since the transmission of these materials varies considerably at short wavelengths. On the other hand, modern synthetic fused silica is effectively clear down to wavelengths below 200 nm, and such cells only need to be matched for the most precise work. The reference cell may seem redundant, particularly in modern instruments with built-in baseline correction, but the correct philosophy is to operate the instrument in the 'null' condition with the baseline as flat as possible. This becomes particularly important at short wavelengths, for in this region the transmission of the cell and solvent are changing rapidly and the sensitivity of the system is falling. Conventional compensation by means of a reference cell is the best method for it is effectively simultaneous with the sample measurement, for digital baseline correction is always subject to temporal effects such as warming-up or drift of the instrument, or thermal effects on the cell or solvent. Some cells − such as flow cells − are too bulky or too expensive to allow an identical reference to be used. A compromise is to use an ordinary cell of the same window material and pathlength as reference. If the sample cell has a reduced aperture, it is desirable to mask the reference beam in the same way.

Cells should never be touched with the fingers or handled by the optical faces. Dry cells can be handled with cotton gloves and wet ones with rubber gloves, but it is better to use plastic forceps, or metal tongs or forceps with their ends protected with short lengths of plastic tubing. The cell should be filled using a glass or plastic pipette. Pasteur pipettes can be used but their volume is too small for a normal 10 mm cell, and for the greatest speed and to avoid the risk of solution entering the bulb it is worth making special transfer pipettes such as that shown in Fig. 9.1. Great care should be taken in filling the cell. The tip of a glass pipette should never be directed towards the windows for these are readily scratched, and for repetitive work it is advisable to protect the end of the pipette with a short length of plastic tube. Of course, when using organic solvents, the compatibility of plastic pipettes or tubing should be checked.

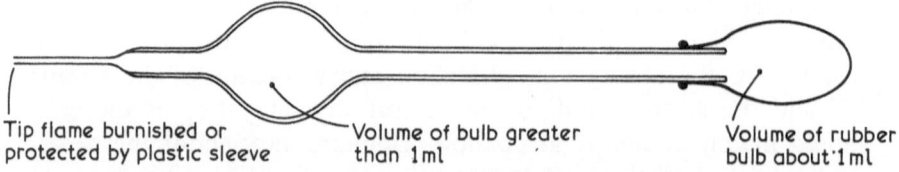

Tip flame burnished or   Volume of bulb greater   Volume of rubber
protected by plastic sleeve   than 1ml   bulb about 1ml

Fig. 9.1 *A specially-made transfer pipette for filling 10 mm normal rectangular cells [1].*

Care must be taken when withdrawing the pipette to ensure that a drop of solution does not run down the outside of the cell. If this contaminates the working area, the only cure when making measurements of the highest accuracy is to empty the cell, clean and dry it and fill it again. In routine circumstances, and if the spillage has not been allowed to evaporate on the cell window, it can be rinsed off with water or solvent and the cell allowed to dry. Again good practice dictates that nothing should touch cell windows: this is imperative for cells used for high-accuracy work or when working in the far-UV. In less demanding situations, the cell can be mopped with lens tissue. Do not rub the cell windows and do not use ordinary laboratory tissues, for these contain abrasive substances and fluorescent brightening agents that will contaminate the windows. Be careful that grease from your fingers is not transferred on to the windows, particularly when using organic solvents, for such contamination can give appreciable absorption and can even cause the appearance of spurious peaks. A serious source of error in spectrometry is the filling and emptying of cells and their relocation in the cell holder. For this reason, the recommended practice is to place the cell in the instrument at the beginning of a series of experiments and not to remove it until the last sample has been measured. The cell holder shown in Fig. 8.8 is ideally suited to this procedure. If the cell is not removed from the instrument, extra care must be taken when filling and emptying it, and if a large number of samples are to be measured, a sipper or flow system will increase speed and accuracy while avoiding the risk of spillage inside the sample housing.

9.4   Making the measurement

Before measuring the sample there are several routine operations that must be performed – these are neglected at your peril, for inquests on poor data can be far more time-consuming than these simple precautions.

(a)   *Warm-up period*
Modern instruments have solid-state electronics that warm up quickly, but tungsten lamps still need 10 min to come to thermal equilibrium and – particularly when old – can give rapid fluctuations during this period. Deuterium discharge lamps are arranged not to strike for 1–2 min after switching on and after this the arc will take some time to settle down.

(b)   *Simple instrument checks*
Ensure that the correct cell holder is in the instrument, that it is properly located and all the cell positions are empty. Check that the correct lamp for the spectral range of interest is alight and that the lamp, filter and detector controls are correctly positioned. On recording instruments, check that there is sufficient paper, that the pens are working and that they are likely to continue to do so.

(c)   *Instrumental baseline*
This can be omitted if the instrument is in regular use over the spectral range to be used. However, the baseline is a useful check of the condition of the whole system and need not take long. Have the cell holders that you intend to use in place but ensure that there are no cells in them and that their apertures are free from dust and incrusta-

tions. Set the instrument operating conditions that will be used for the sample and scan fairly quickly over the range of interest. Even if measurements are to be made at a single wavelength, it is worth scanning on each side of that wavelength to ensure that the baseline is not changing rapidly. The baseline should be smooth, correspond to 1.0 T or 0 A and be free from excessive noise. If these conditions are not met, consult Section 9.5 or the more elaborate checks described in Chapter 11.

(d)   *Cell baseline*

This should be performed every time that clean cells are put into the instrument, and at least once per day. For the most accurate work, it is essential to run the cell baseline on each piece of chart paper that is used. The sample and reference cells are placed in position and filled with the appropriate solvent. When using cells with a restricted working area such as micro-cells, check that the beam passes through the cell without touching the cell walls. This check can be done by turning the wavelength to 550 nm, opening the slit, and in subdued lighting or with a piece of black cloth over the open cell compartment, trace the path of the beam through the cell holder and cell by means of a small piece of white card. With micro-cells some loss of light is inevitable but it is essential that this falls on the mask and that there is no risk of it passing through the cell walls or over the sample. The best micro-cell holders are provided with some means of adjusting the position of the cell laterally. Failing this, it may be possible to adjust the position of the cell holder in the instrument so that the cell is centrally placed in the beam. If there is any risk of movement of the cell in the holder it can be wedged against the reference edge or corner with pieces of card or thin metal.

The vertical alignment of the beam should also be checked with a known volume of water in the cell. This was illustrated in Fig. 8.2. The bottom of the beam should be above the cell floor and the top of the beam should fall at least 1 mm below the lowest part of the meniscus. If a limited amount of solution is available, the cell should be raised in the holder by packing pieces of card or plastic to minimize the clearance between the beam and the cell floor. If the sample volume is insufficient to fill the cell to the proper level then the vertical height of the beam must be reduced with a mask.

Having positioned and filled the cell or cells with solvent, scan through the wavelength range. Since no peaks are expected in this trace, it can be scanned fairly rapidly. If the cells and the solvent they contain are identical, this trace should coincide with the instrumental baseline. In practice, this is rarely the case, particularly at short wavelengths. When approaching the shortwave solvent cut-off and the instrument is 'running out of energy' as a result, deviations from the expected baseline may result from the pen running off scale or going 'dead'. Otherwise, if there are considerable differences from the instrumental baseline, check that the cells look clean and that there are not particles suspended in the solvent. Exchange the cells and see whether the deviation is reversed in sign. Empty and refill the cells with fresh solvent and re-measure. If the problem remains, clean the cells again and fill them with freshly distilled water. If they still show imbalance, this may be due to inherent differences in the transmission of the windows, contamination that your cleaning procedure has failed to remove, or fine scratches that are not visible when the cell is filled. If it is decided to procede with the measurement using these cells, make regular checks of the solvent baseline to ensure that it is not changing during the day due, say, to the dissolution of contaminants from the windows.

(e)   *Measuring the sample*

The solvent is drawn out of the sample cell and, if sufficient sample solution is available, a few drops are used to rinse the cell. It is then filled to the correct level. It may not be possible to fill and empty short-pathlength cells while they are in the instrument. The tip of a standard Pasteur pipette is liable to jam inside a 1 mm normal cell and may scratch the windows or break off inside the cell. It is better to draw your own pipettes from tubing with a thicker wall so that the outside diameter is about 0.5 mm, and flame-burnish the end.

Before running the spectrum, a choice of operating conditions must be made. The number of operating parameters that can be chosen by the operator varies between instruments, but the aim should be to scan the selected wavelength range as fast as possible without any loss of accuracy of the record, both in wavelength and absorbance terms. If the scan is made faster than the data capture system can deal with it, maxima will appear displaced to longer wavelengths, be reduced in height and distorted. Selection of the scanning speed is largely a matter of experience based on the performance of your instrument over a particular wavelength range and when measuring a spectrum of the same complexity. If the spectrum has sharp peaks with fine structure, a slow speed must be used. If you are uncertain about accuracy of the resulting spectrum, scan it again at a slower speed and compare the results. If the peaks are unchanged in position and shape, the original speed was satisfactory. Fig. 9.2 shows the result of scanning the benzene vapour spectrum — which has very sharp peaks and more fine structure than is seen in most solution spectra — at different speeds with a conventional recording spectrometer. An ideal solution to the scanning speed problem is presented in some modern microprocessor-controlled spectrometers where the scanning speed is automatically regulated by the rate of change of the absorbance, the instrument scanning fast over flat regions of the spectrum and slowly over sharp peaks.

While most instruments allow a choice of recording speed, far fewer permit a choice of slitwidth (ESW). Here the compromise is to select the smallest ESW and hence greatest spectral resolution consistent with reasonable sensitivity. In prism instruments, the ESW for a given slit setting varies with wavelength due to the changing dispersion of quartz. This is illustrated in Fig. 9.3. When using manual instruments with quartz prisms, the operator must select a slitwidth for each wavelength, the choice being based on the ESW, the lamp output and the detector response at that wavelength. The manufacturer's manual generally gives the slit settings to maintain constant sensitivity. In most recording prism instruments, the slit servo operates to maintain a constant response from the detector and hence a roughly constant sensitivity at all wavelengths. This means that there are considerable fluctuations in the ESW as the instrument scans, as shown by the broken line in Fig. 9.4. In critical applications, in order to obtain a specific ESW at a given wavelength it may be necessary to override the slit mechanism and set the slit by hand.

The problem is simpler in grating instruments for in these the dispersion is constant and so the instrument has constant ESW with a fixed slitwidth. It is still necessary for the instrument to compensate for changes in the lamp output and detector sensitivity with wavelength, but this is usually done with an optical wedge or comb that adjusts the amount of radiation without changing the ESW.

As a rule of thumb, the ESW should be not more than of the width at half-height of the sharpest band to be measured or, alternatively, the ESW should be less than one-eighth of the bandwidth: thus for a peak of bandwidth 40 nm, the ESW should not exceed 5 nm. Fig. 9.5 demonstrates the effect of bandwidth on the apparent height of a sharp peak in the benzene spectrum, but significant deviations can occur with relatively broad peaks if excessive ESW settings are used.

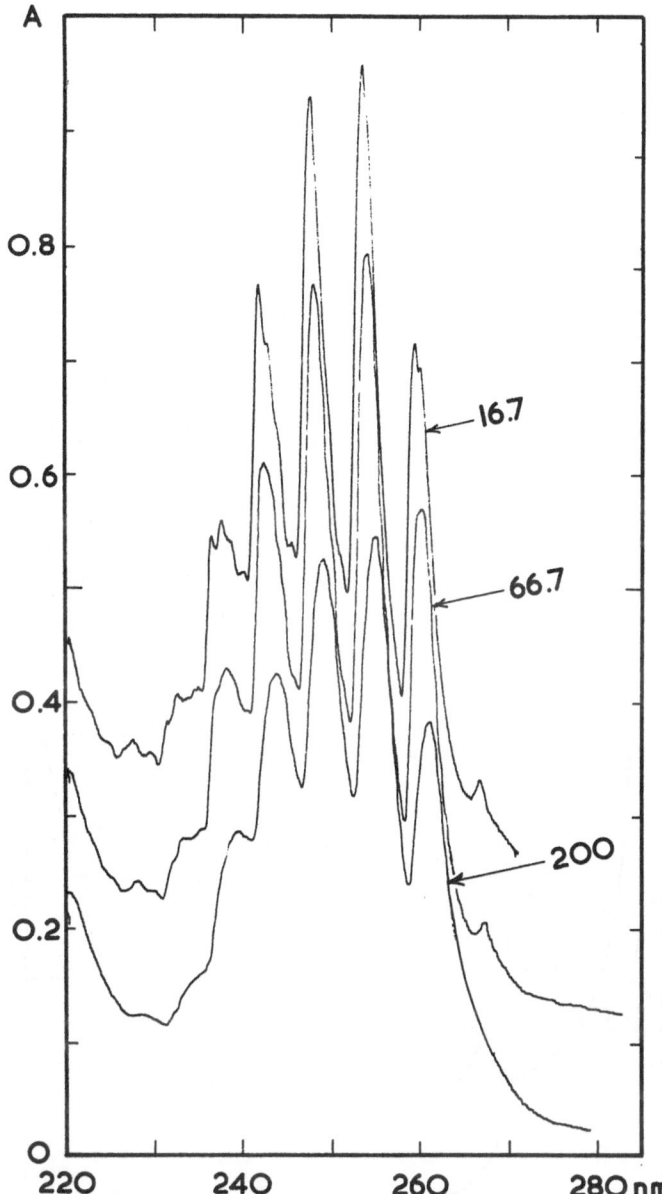

Fig. 9.2 *Effect of scanning speed on a recorded spectrum. A cell containing benzene vapour was scanned from short to long wavelength at three different rates given as nm min^{-1} on a Shimadzu MPS-5OL double-beam spectrometer. Excessive speed has reduced the fine structure, lowered the peaks and raised the minima, and has shifted both maxima and minima to longer wavelengths. These effects are due primarily to the slowness of the pen response.*

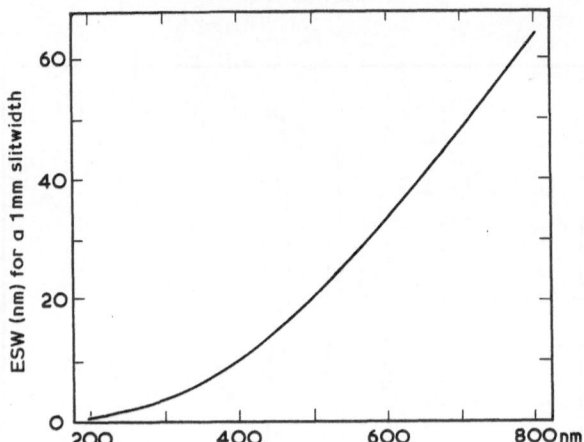

Fig. 9.3 *A typical dispersion curve for a quartz prism monochromator. This shows the effective spectral slitwidth at different wavelengths for a slit opening of 1 nm.*

If you are uncertain about your choice of ESW, scan the critical peak again with a smaller slit setting. Even if the slitwidth of your instrument cannot be changed, it is essential to know the ESW at different wavelengths and appreciate that a potential problem exists.

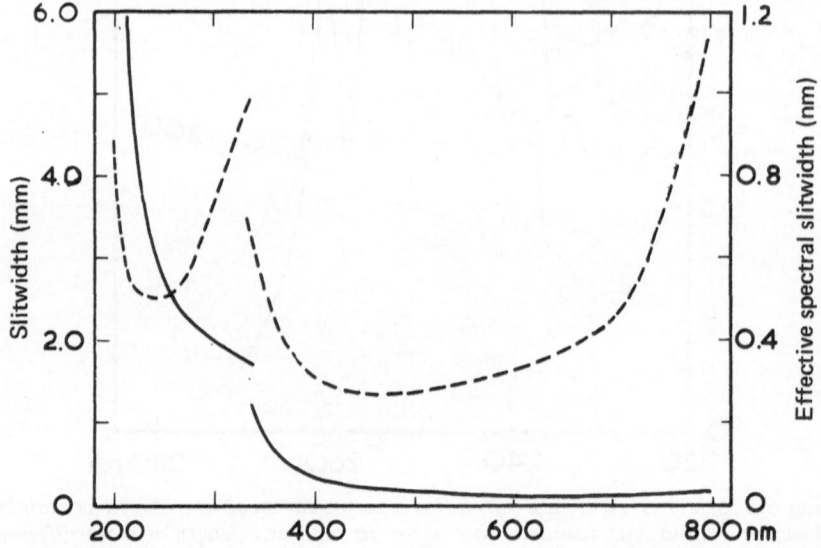

Fig. 9.4 *Automatic slit settings for the Shimadzu MPS-5OL under normal operating conditions (continuous line) and the resulting ESW values calculated from Fig. 9.3 (broken line).*

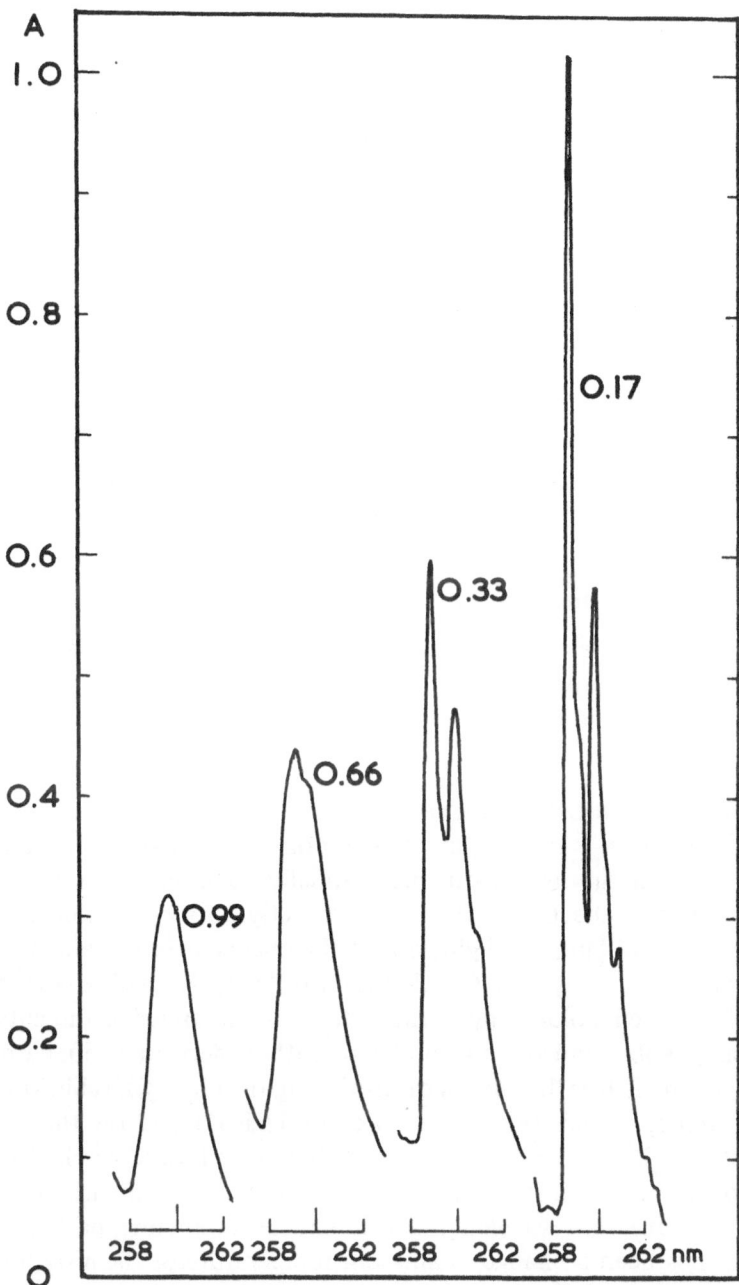

Fig. 9.5 *Distortion of the recorded absorption spectrum due to excessive ESW. A cell containing benzene vapour was scanned at the same rate with four different ESW values, which are given in nm. The spectrum at the smallest slit setting approaches the true shape of the band, the peaks of which are extremely narrow.*

The instrument is then set to the starting wavelength by driving the mechanism in the direction of the intended scan so that the backlash in the driving mechanism is taken up. If possible the pen should be allowed to equilibrate before starting the scan to avoid generating spurious peaks. If the peak absorbance of the sample lies outside the chosen absorbance range, it is generally better to complete the scan and then go back and fill in the missing parts of the spectrum on another scale. If the critical part of the spectrum is found to lie above the recommended range, i.e. 1.5 A, consider diluting the sample or changing to a shorter pathlength cell.

At the end of a series of measurements, the cells should be removed from the instrument, emptied and submerged in solvent or one of the cleaning solutions suggested in Section 9.6. Cells should never be allowed to dry out as this makes cleaning far more difficult.

9.5   Problems and pitfalls

Experience will teach when a spectrum does not 'look right' even though the above procedures have been carried out. Some commonplace problems will be described.

(a) *Poor instrumental baseline*

Check that there is nothing in any cell holder, that they are correctly aligned and, if the instrument has manually operated filters, shutters or attenuators, check that these are correctly positioned. Check that the appropriate lamp is alight, and if the change-over is manual, that the lever is properly set. Check that the beam is passing correctly though the cell holders by turning to 550 nm, opening the slits as wide as possible and tracing the beam with a piece of white card. If the instrument has the means of displaying the energy level in single-beam mode, check this: otherwise, excessive noise on the trace suggests that lamp output is low or that there is failure of an electronic component. If the instrument has some means of compensating the baseline at different wavelengths, this may need resetting, particularly if there has been a change in ambient temperature or the instrument has not been used for some time. If the baseline has sharp spikes or steps that always occur at the same wavelength, suspect some fault in the automatic filter-changing system. Further checks are given in Chapter 13.

(b) *Poor cell baseline*

If the instrumental baseline is satisfactory, then the fault must lie with the cells or their contents. Empty and refill the cells to ensure that the solvent is homogeneous. Visually check the cells for dirt inside and for streaks or scratches on the windows. If using solvents other than pure water, empty the cells and refill with water, for high solvent absorption at short wavelengths may show up a latent imbalance in the instrument or differing stray-light levels in the two beams. If the problem is still not resolved, try another pair of cells or clean the cells again using a more drastic method.

(c) *Low absorbance values*

When measuring the sample, the absorption maxima may appear lower than expected for a number of reasons:

(i) Incomplete solution — check that there is no residual material in the flask and no suspended particles in the solution. Check that the solute has not come out of solution in the cell as a precipitate or as a coating on the walls.

(ii) Sample impurity — it could be contaminated with a non-absorbing impurity.

(iii) Changes in the sample — Check that the pH of the solution is correct. Could the sample have suffered photochemical or oxidative damage? Check the rest of the spectrum to see whether there has been a corresponding absorbance increase due to product formation, and run the spectrum again to see whether the change is a progressive one.

(iv) Bubbles in sample or reference cells — if there have been temperature changes or the cells have been filled for some time, bubbles may have formed on the windows.

(v) Light by-passing the sample — Check that the beam cannot pass above, below or either side of the sample, and that there are no bubbles in the cell.

(vi) Peak distortion due to excessive scanning speed or too great an ESW — this is only likely to occur with very sharp peaks.

(vii) When making measurements at short wavelengths always be alert to the contribution of stray-light. This is equivalent to light by-passing the sample and so will reduce the apparent absorbance (see Fig. 9.6). The problems can be improved by diluting the sample or using a shorter pathlength cell so that the light transmitted by the sample is increased and the

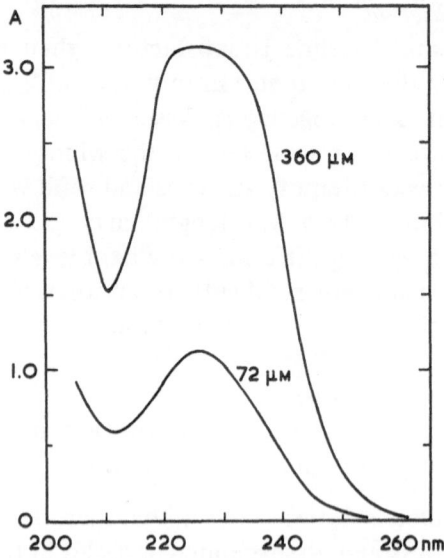

Fig. 9.6 *Distortion of a recorded absorption spectrum due to stray-light. Two aqueous solutions of potassium iodide differing in concentration by a factor of 5 were measured in a 10 mm cell on a Pye Unicam SP8–100. The curve at low concentration represents the true spectrum while that for the high concentration is distorted below 240 nm. The apparent* λ_{max} *is shifted and is only about 60% of its true height.*

stray-light is a smaller proportion of the light reaching the detector.

(viii) Beer's Law failure – the linear relationship of absorbance with concentration may appear to fail as a result of a change in the molecular species in solution due to dimerization, complex formation etc. If such a problem is suspected, measure a series of dilutions and see whether the change is progressive. While Beer's Law failure is often invoked to explain non-linearity, in the majority of cases the error has some other cause.

(ix) Fluorescence – there is a possibility that the measuring beam may cause the sample or solvent to fluoresce: if this emission enters the detector it will reduce the apparent absorbance. This is unlikely to cause serious error.

If none of the above suggestions seems to explain your problem it will be necessary to make a proper check of the absorbance accuracy of the instrument using the methods detailed in Chapter 13.

(d) *High absorbances*

This is a far less common problem, and so in double-beam instruments it is worth first considering whether the transmission of the reference beam has increased for any reason. Otherwise the fault is probably optical obstruction of the sample beam by bubbles, evaporation of the solution increasing its absorbance or causing the meniscus to descend into the beam, or the formation of a precipitate in the solution. When making measurements at temperatures below ambient, be alert to the possibility of condensation forming on the cell windows, the solute coming out of solution, or the solution freezing, all of which will increase the apparent absorbance.

(e) *Incorrect λ_{max} values*

Check that the wavelength scale on the chart corresponds to that on the monochromator. In critical applications superimpose the transmission profile of a calibrating filter or solution on the same chart: remember that chart paper will expand with increases in humidity as the day goes on, and the wavelength calibration of the instrument can drift as it warms up.

When measuring mixtures, the apparent λ_{max} of minor components appearing on the side of major peaks will be displaced from the true value. The effect of the superposition of a minor peak on the tail of a major one is illustrated in Fig. 9.7a. If there is overlap of two peaks of similar size, the resulting single peak gives no clue as to the presence of two components (Fig. 9.7b). Changes of pH or solvent can give major differences in λ_{max} from published values and, when this information is available, it is worth checking that the concentration and temperature of the solution are similar to those of the reference. As we have seen, stray-light can cause spurious peaks to appear at short wavelengths and can also displace the position of true peaks. If the peaks are flattened and have an asymmetrical appearance, stray-light is probably present.

(f) *Spurious peaks*

Unexpected peaks that are not present in the solvent baseline can be the result of impurities or degradation products in the sample or due to contamination introduced during the handling of the solution.

9.6 Cell cleaning

Cell cleaning procedures were dealt with in Volume 1 of this series [1] and will be only briefly dealt with here. The first principle is to

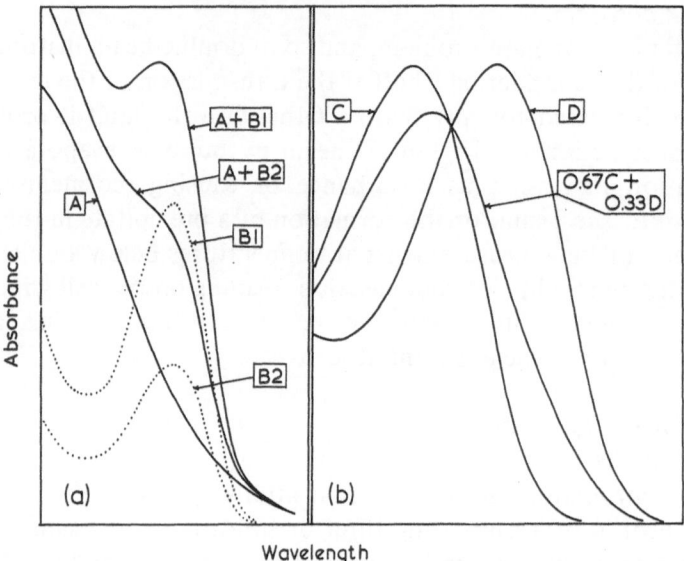

Fig. 9.7 *Diagram illustrating the distortion of spectra by the overlap of bands: (a) the apparent λ_{max} of the minor band is displaced towards the maximum of the major peak; (b) a 2:1 mixture of compounds with overlapping spectra results in a single peak of intermediate λ_{max}.*

avoid drastic cleaning methods if possible, for they are time consuming and potentially dangerous to both cell and user. Many cleaning problems can be avoided by not leaving solutions in cells for longer than necessary and *never* allowing them to evaporate.

When cleaning a cell first assess the problem. Is there a suitable solvent for the contaminant that will not damage the cell? Plastic cells can only be cleaned with dilute solutions of detergents. Glass cells must not be treated with alkaline solutions and there is a possibility that cheaper cells may be of cemented construction which will exclude the use of strong acids and alkalis and some organic solvents. Silica cells of fused construction are effectively inert, and can if necessary be heated in the cleaning solution.

Organic contaminants may well be dissolved by a suitable organic solvent. Failing this, a cold or warm solution of a detergent may be successful. The next approach is to oxidize the offending substance with chromic acid mixture or hot nitric acid. Extreme care must be taken when handling these reagents: this should be done in a fume hood wearing gloves, goggles and protective clothing. If oxidizing agents are used it is essential to make sure that the cell is as free of organic matter as possible; first rinse it with acetone or ethanol and then several times with water.

Finally the cell should be thoroughly rinsed with distilled water. If it is in regular use with aqueous solutions it is best to store it submerged in water, upright in a narrow container so that it cannot fall over. If it is necessary to dry the cell, empty it and stand it inverted on an absorbent surface. A small plastic-coated test-tube rack can be used to prevent the cell falling over, but this must not touch the windows. To dry the cell more quickly it may be rinsed with spectroscopic grade alcohol or acetone; this should be drained from the cell as completely as possible before allowing it to evaporate.

Cells should never be touched with the fingers especially when wet. When a routine of good cell handling has been established, it takes no longer than more casual methods, guarantees the best possible results and avoids time-wasting inquests on poor results.

Dry cells should be stored in small sealed boxes free of dust and lint. Plastic sandwich boxes are ideal, but the plush-lined boxes supplied by some manufacturers should be discarded. For transportation or long-term storage the cells may be wrapped in lens tissue.

9.7   Accuracy and precision in absorbance measurement

Your sample has been measured with as much care as possible and your instrument has produced an absorbance value at a particular wavelength. How valid is this number? Any measuring system has two criteria of performance, precision and accuracy. These are illustrated in Fig. 9.8 by the performance of a marksman in a rifle range. In (a) the shots are widely distributed all over the target – this shows neither precision nor accuracy. In (b) they are grouped together away from the bull's-eye: the marksman and rifle are performing well

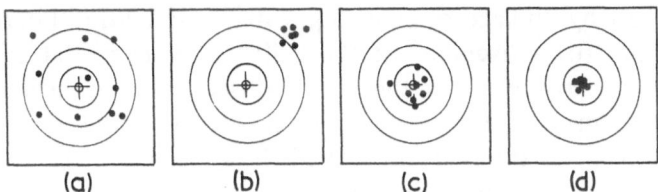

(a)   (b)   (c)   (d)

Fig. 9.8 *The concepts of precision and accuracy are illustrated by the performance of a marksman firing at a target: (a) lacks both precision and accuracy; (b) the hits are precisely located but are inaccurate; (c) here the accuracy is good but the precision poor; (d) shows both precision and accuracy. From Altman [2].*

and giving a precise distribution of shots, but there is a bias — perhaps the rifle sight is bent. Target (c) shows a grouping with satisfactory accuracy but poor precision: the average of a large number of hits would lie on the bull's-eye. In (d) the hits are both precise and accurate. The analogy with absorbance measurement is good: both instrument and operator may be working reliably to give precise data and repeated measurements on the same sample show small deviation, but some error in sample preparation or instrumental bias means that the mean of the measurements does not represent the true absorbance. To obtain results that are both precise and accurate demands precision and accuracy from both operator and instrument. Precision is checked quite easily by repeated measurements. Modern instruments are capable of extremely high precision and in 99% of cases it is operator error that leads to poor data. On the other hand, assessment of accuracy is far more difficult, and in laboratories where quantitative results are vital, a great deal of effort must be put into checking and improving the accuracy of absorbance measurements.

9.7.1   Tests for precision

Suppose that you are involved in the routine measurement of a number of solutions of the same solute. The first test would be to take a typical solution and perform your normal routine of filling, measuring and emptying the cell, say, 20 times. Calculate the standard error of these results:

$$s = \left(\frac{\Sigma x^2 - (\Sigma x)^2 / n}{n - 1} \right)^{1/2} \tag{9.1}$$

The standard error will depend upon the instrument and the nature of the sample, but should be of the order of 0.005 and not exceed 0.02. If it does, consider the following potential causes of deviation: (a) inhomogeneity of sample; (b) contamination of the external faces of the cell by spillage or fingering; (c) contamination internally by the solute; (d) movement of the cell; (e) instrumental instability; (f) solvent carry-over — if the cell is rinsed between measurements, the sample may be diluted by residual solvent. If the test is satisfactory, next assess the performance of the system over the anticipated range of absorbance readings.

9.7.2   Linearity checks

Prepare a series of dilutions of the most concentrated sample and measure their absorbance using your normal routine. Plot these values against concentration. Ideally it should be linear over the whole range with the points all falling on the line. The line may deviate from linearity due to stray-light, instrumental non-linearity or Beer's Law failure, but most probably due to poor volumetric technique. Sample handling errors can probably be diagnosed by repetitions of the measuring procedure, but instrumental and Beer's Law problems are difficult to distinguish. If a solution is measured in cells of different pathlength, then any non-linearity in the absorbance—pathlength relationship must arise in the instrument. On the other hand, if solutions of differing absorbance are measured in the same cell, any non-linearity may be due to Beer's Law failure, instrumental non-linearity, stray-light, or some combination of these effects. As Clarke [3] points out, a test of this sort may underestimate the Beer's Law deviation.

Scatter of the points may be due to any of the causes outlined in the previous section. If the scatter becomes worse at high or low absorbances then it may be instrumental fluctuations that are limiting the precision of the results. In any case, it is prudent next to consider in what ways the precision of the measurements can be improved.

9.7.3   *Optimization procedures*

Having established that the sample handling is being performed as carefully as possible and that the instrument is working well, precision can be further improved in two ways: first, by increasing the number of measurements and thus reducing the standard error, and second, by ensuring that measurements are made within the optimum range of the instrument. Repeated measurements will only increase precision if *all* the critical steps in the measurement are repeated: if cell filling is the prime cause of deviation, then the cell must be filled and emptied for each of the replicates. Selection of an optimum range is more concerned with the performance of the instrument. Fig. 9.9 shows the instrumental error in the measurement of a series of solutions of a given solute at a specific wavelength on a particular instrument. The curves represent the deviations in the absorbance values when each of the solutions was measured a number of times. The 0 and 1.0 T of the instrument was reset before each measurement, but the sample cell was left in the cell holder. As expected, the precision falls at high and low absorbancies, while the precision is increased the number of repeated measurements. The optimum absorbance range

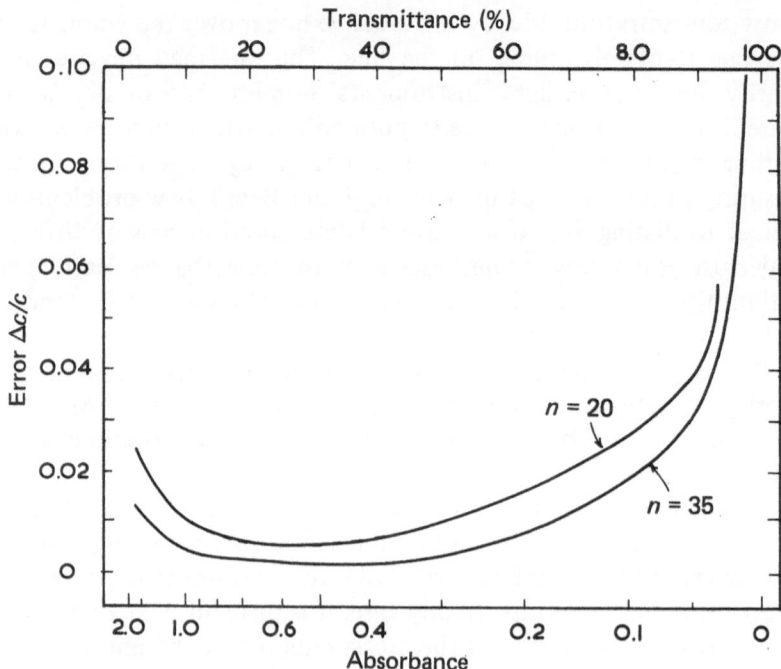

Fig. 9.9 *Measurement of instrumental error in a spectrometer. The transmission of a series of 12 solutions of potassium chromate measured at 370 nm with a Bausch and Lomb Spectronic 20. Each solution was measured 20 and 35 times, and error values based on the deviation of these points were calculated by Equation (9.2). From Youmans and Brown [7].*

for a given level of precision can thus be increased by increasing the number of repetitions.

There are a number of intrinsic causes of instrumental deviation, the most important being noise generated by the detectors ('shot' noise), noise generated by the electronic components of the amplifiers, fluctuations of the source and 'noise' generated by the uncertainty of the read-out. The contribution of these various kinds of noise to the signal-to-noise ratio is discussed in detail by Rothman *et al.* [4] who make recommendations on the selection of operating parameters to maximize the signal-to-noise ratio.

An early attempt to establish the optimum absorbance range of the Beckman DU spectrometer was made by Vandenbelt *et al.* [5] who measured the molar absorptivities of various compounds at different concentrations. Three typical curves are given in Fig. 9.10, which shows the deviations of the apparent molar absorptivity when measured with solutions of high and low absorbance. The linear part

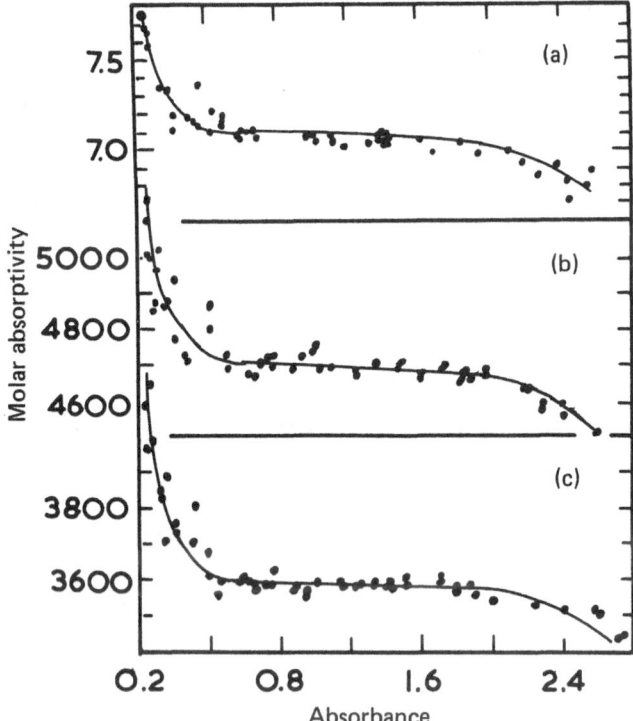

Fig. 9.10 *Non-linearity in a manual spectrophotometer. A plot of the variation in the apparent molar absorptivity of two compounds at three wavelengths with absorbance of the test solution. (a) Potassium nitrate in water at 301 nm; (b) potassium chromate in 50 mM potassium hydroxide at 373 nm; and (c) potassium chromate in 50 mM potassium hydroxide at 273 nm. Redrawn from Vandenbelt et al. [5].*

of the plot indicates the absorbance range of greatest accuracy. Interestingly, all three curves are of the same shape although they were measured at different wavelengths and correspond to widely differing solute concentrations. The downward trend in molar absorptivity values above 2 A is probably due to stray-light, but the sharp rise in the values at low absorbance is less readily explained. The authors suggest that it may have been due to photometric non-linearity in the system.

An alternative approach to the determination of the optimum concentration range for a spectrometric assay was proposed by Ayres [6] based on an earlier proposal by Ringbom. Rather than plotting absorbance against concentration for a series of test samples, transmission is plotted against the logarithm of their concentration (Fig.

9.11). The linear region of maximum slope indicates the range of concentration that will give maximum sensitivity and linearity for the determination. The method is based on the assumption that instrumental error in making transmission measurements is constant throughout the transmittance range, which is not true. A more rigorous method for estimating the region of maximum precision of an instrument is given by Youmans and Brown [7]. The error in the measurement of the concentration of a solution of transmission T can be expressed as:

$$\Delta c = \frac{1}{eb} \log \frac{T_{av}}{T_{min}}$$

where e is the molar absorptivity of the compound, b the pathlength

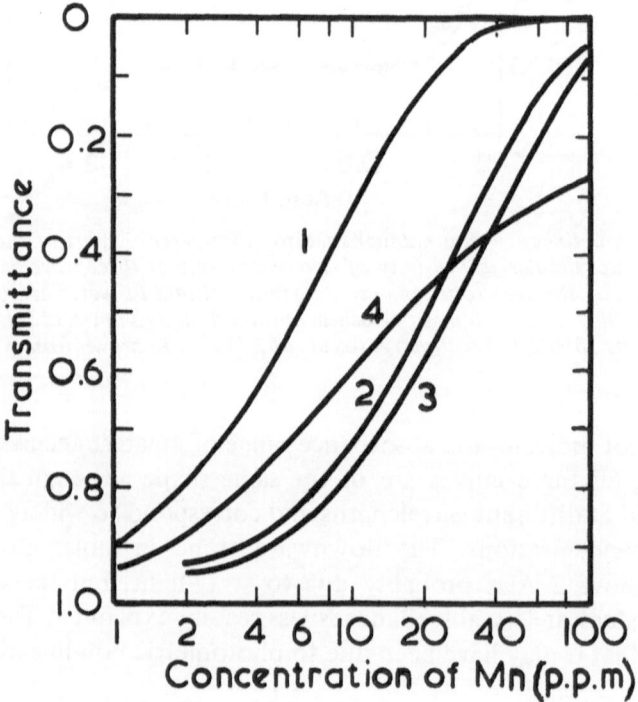

Fig. 9.11 *An Ayres plot for the determination of manganese as permanganate. Curves 1–3 were obtained with a series of solutions measured at 526, 480 and 580 nm, respectively, using a Beckman DU manual spectrometer. Curve 4 was obtained using a filter colorimeter. Instrumental accuracy is greatest where the slope of the curve is greatest, deviations occurring at high and low transmittances. Redrawn from Ayres [6].*

and T_{av} and T_{min} are the mean and minimum values from a series of n measurements on the same sample.

The statistical error is then:

$$\frac{\Delta c}{c} = \frac{1}{-\log T_{av}} \log \frac{T_{av}}{T_{min}}$$

The standard error s of T_{av} for the population of n measurements can be calculated from Equation (9.1) and corrected for bias with Student's t-value. Then

$$T_{min} = T_{av} - \frac{ts}{n}$$

In this way, the curves shown in Fig. 9.9 were calculated from groups of 20 and 35 measurements on each of a series of 12 solutions covering the transmission range. Values of T_{av} and s were calculated and then T_{min} derived using Student's t-values taken from tables taking the degree of freedom as $(n - 1)$ and $\alpha = 0.05$.

These curves, of course, refer to one particular type of estimation on a specific instrument, but are fairly typical. They show the relative intrinsic error in the system at different absorbance values. They do not form any basis for the correction of such errors but are of great help in the optimization of a particular assay.

9.8   Difficult samples

9.8.1   *Solid samples*

Satisfactory absorption measurements can be made on samples that are insoluble or must not be dissolved, though the techniques are more difficult and the results more liable to mis-interpretation. For large optically homogeneous specimens such as single crystals, glasses or polymer sheets and blocks, satisfactory measurements can be made simply by fixing the sample in the measuring beam, with suitable masks if necessary. For the best results, the sample should have parallel plane faces and be arranged so that these are perpendicular to the beam. For irregularly shaped specimens, immersion in a liquid of matching refractive index can reduce the effects of surface reflection and refraction. Of course, the liquid must not absorb appreciably in the region of interest.

Smaller crystals can be measured with a micro-beam set-up (see Section 9.8.3) or by packing a quantity of them into a cell. Light-scattering from the surfaces of the crystals will be a major problem, though the effects of this can be reduced (see Section 9.8.2). It can also be reduced by suspending the sample in a high-refractive index liquid to form a 'mull'. Everett [8] has measured insoluble compounds in the UV by mixing them with a non-absorbing salt and compressing the mixture into discs, just as in the IR 'pressed disc' technique. Light scatter is a major problem but can be minimized by optimizing the size of the salt granules.

For low-melting point solids it is worth considering measuring them in the molten state: temperatures up to 100°C can be maintained in a suitably insulated cell holder in a conventional spectrometer. The results should be interpreted with caution for the liquid state spectrum may not be identical with that in the solid state or in solution. An alternative technique, which is the only one that can be used for highly absorbing solids, is reflectance spectrometry.

9.8.2   *Scattering samples*

A common problem, particularly with biological samples, is that the substance of interest is associated with suspended particles of other material. Although the latter may not absorb in the spectral region of interest, scattering of the measuring beam will distort the measured spectrum and may raise the apparent absorbance of the sample above the optimum range of the instrument. The scattering of radiation depends upon the size of the particles involved and is also a power function of the frequency of the radiation – that is, its effect increases rapidly when scanning down into the UV region. This is illustrated by Fig. 9.12 which shows the effect of increased turbidity upon an absorption spectrum. Scatter increases the apparent absorbance, particularly at shorter wavelengths, and so the spectrum is distorted and its peaks are shifted to shorter wavelengths. Keilin and Hartree [9] presented a mathematical treatment of this effect.

Some degree of compensation can be made by using a scattering suspension in the reference beam – diluted milk or 'Ludox' colloidal silica may be helpful. Goldbloom and Brown [10] give an example of this. A better solution is to place the sample cell nearer to the detector – the latter then receives a greater proportion of the scattered radiation (Fig. 9.13) and the measured spectrum will be nearer the true shape. Some modern instruments have an alternative cell holder placed in front of the detector to facilitate this. An alternative method for use with conventional spectrometers was proposed by Shibata *et al.* [11] who showed that if a scattering 'screen' is placed between the cell and the detector, the absorption spectrum will be nearer its

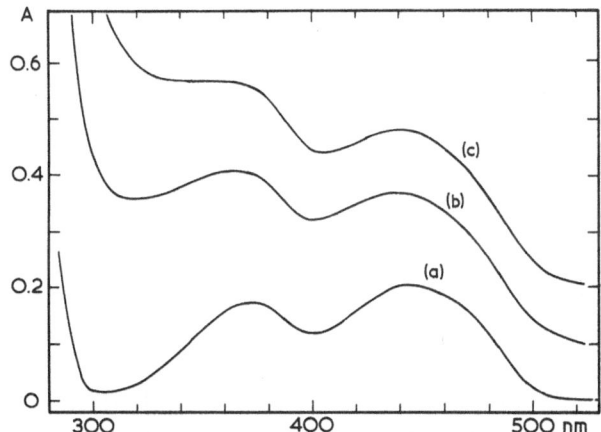

Fig. 9.12 *The effect of light-scattering by the sample, and instrumental compensation: (a) absorption spectrum of 16.9 μM riboflavin in water at pH 5.6 measured in a 10 mm cell with a Pye Unicam SP8–100; (b) the same solution with 1% milk added measured in the 'turbid sample' position, close to the detector; (c) the solution containing milk measured in the normal cell position.*

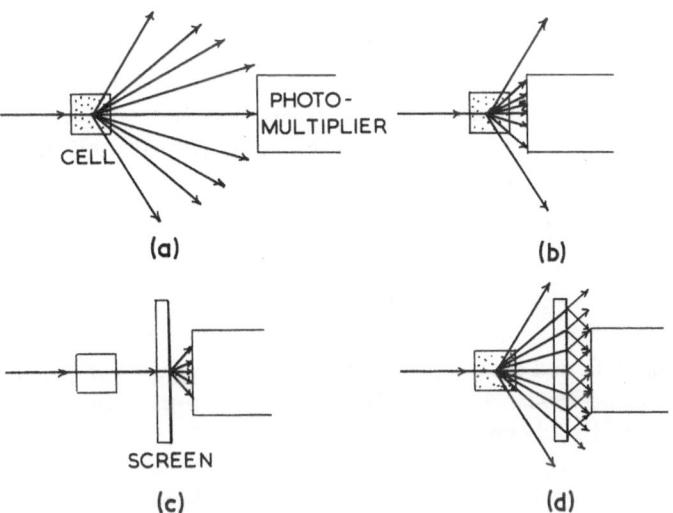

Fig. 9.13 *Instrumental methods for correcting for light-scattering. (a) Conventional arrangement of cell and detector. A scattering sample in the cell causes much of the radiation leaving the cell to miss the detector; (b) if the cell is placed close to the detector, a larger proportion of the radiation will enter it; (c) if a diffusing screen is placed between sample and detector, light passing straight through the sample will be scattered before entering the detector; (d) a scattering sample will show the same transmittance as a clear one since the same proportion of light will enter the detector irrespective of how it strikes the screen.*

true shape. All light rays emerging from the cell will be scattered equally by the screen irrespective of their direction, and so a constant fraction of the light striking the screen will enter the detector. Shibata *et al.* used oiled paper as a screen, but 'flashed' opal glass is more satisfactory.

9.8.3   *Microscopic samples*

The measurement of microscopic samples is becoming commonplace and many special instruments have been built for the measurement of forensic samples, chromosomes in cells, and so on. Accessories are available for conventional spectrometers that will reduce the beam to microscopic size but probably custom-built instruments based on a microscope are more satisfactory. Commercial instruments of this type are reviewed by Altman [2] and can be very expensive. Simpler home-built instruments designed for a particular task can give equally good results at far less cost. The instrument described by Liebman and Entine [12] is still in use in up-dated form and produces excellent absorption spectra in the range 320–700 nm from specimens 0.5 x 2 μm in size. The major problem in making quantitative measurements with this type of instrument is to provide a satisfactory reference. In Liebman's instrument a reference beam is taken through the microscope and focused on the specimen slide beside the sample beam. The operator manoeuvres the slide so that the sample beam passes through the sample while the reference passes through the mounting medium.

References

1   Burgess, C. and Knowles, A. (Eds) (1981), *Techniques in Visible and Ultraviolet Spectrometry*, Vol. 1, *Standards in Absorption Spectrometry*, Chapman and Hall, London.
2   Altman, F.P. (1981), *UV Spectrom. Grp Bull.*, **9**, 28.
3   Clarke, F.J.J. (1981), *UV Spectrom. Grp Bull.*, **9**, 81.
4   Rothman, L.D., Crouch, S.R. and Ingle, J.D. (1975), *Anal. Chem.*, **47**, 1226.
5   Vandenbelt, J.M., Forsyth, J. and Garrett, A. (1945), *Anal. Chem.*, **17**, 235.
6   Ayres, G. (1949), *Anal. Chem.*, **21**, 652.
7   Youmans, H.L. and Brown, V.H. (1976), *Anal. Chem.*, **48**, 1152.
8   Everett, A.J. (1968), personal communication.
9   Keilin, D. and Hartree, E.F. (1958), *Biochim. Biophys. Acta*, **27**, 173.

10 Goldbloom, D.E. and Brown, W.D. (1966), *Biochim. Biophys. Acta*, **112**, 584.

11 Shibata, K., Benson, A.A. and Calvin, M. (1954), *Biochim. Biophys. Acta*, **15**, 461.

12 Liebman, P.A. and Entine, G. (1964), *J.Opt. Soc. Amer.*, **54**, 1451.

10 Numerical methods of data analysis

10.1 Baseline corrections

10.1.1 *Irrelevant absorption*

The measured absorbance at any wavelength is the sum of the absorbances of the various components present and this is the basis of the quantitative analytical methods, discussed in Section 10.2. This approach presupposes that all of the absorbing components present are known, or can be identified. However, this is frequently not the case particularly with measurements on biological materials. There is background absorption, commonly termed *irrelevant absorption*, which is usually rather non-specific in that it lacks discrete maxima. If this absorption can be quantified and subtracted the spectra may be used with confidence for assay work. Various methods have been described for the elimination of this irrelevant absorption, some very simple and others more mathematically demanding. The selection of a particular method is very dependent on the nature of the irrelevant absorption; for example, if it is linear, two points in principle, serve to define it. The earlier work on baseline correction procedures has been well reviewed by Mulder *et al*. [1], whose paper may be consulted to supplement the basic principles given below.

10.1.2 *The Morton and Stubbs correction*

The simplest approach to background absorption is to assume that it increases linearly below the peak of the absorption band being studied at the same rate as on the long wavelength side of the band. The weakness of this approach is that it usually involves a long extrapolation of the baseline and the assumption of linearity is then often not valid. It is therefore better to use a procedure which fixes the baseline over a short distance only on either side of the absorption maximum; this is the basis of the Morton and Stubbs method [2], which is demonstrated by Fig. 10.1. The upper curve is the measured absorption band, the lower broken curve is the true contour of the compound to be estimated and the background is assumed to be linear over the small wavelength range under consideration. Three wavelengths, λ_1, λ_2 and λ_3 are chosen, such that $\lambda_2 = \lambda_{max}$ and λ_1

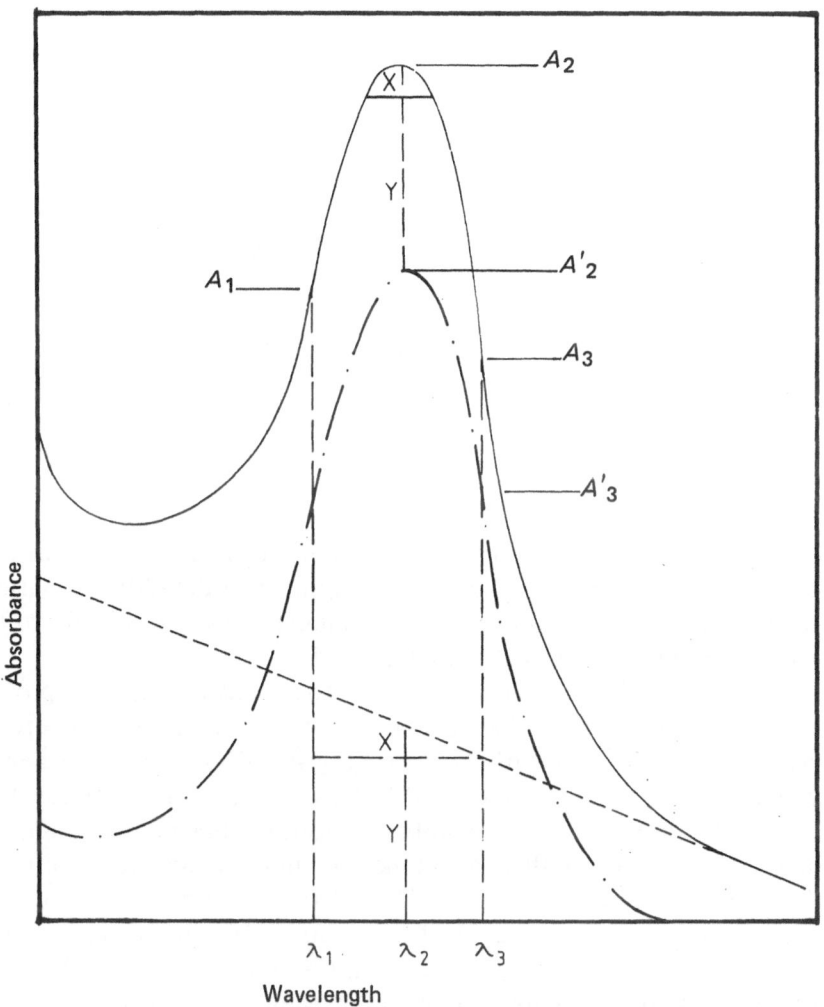

Fig. 10.1 *The Morton and Stubbs baseline correction. A_1, A_2 and A_3 are the absorbances at three points on the observed absorption spectrum (——). A'_2 is the absorbance at λ_2 on the true absorption spectrum (—·—·—) after correcting for the background absorption (———) which is assumed to be linear between λ_1 and λ_3. Redrawn from Morton and Stubbs [2].*

and λ_3 have equal molar absorptivities, i.e. $\epsilon_1 = \epsilon_3$, which are a convenient, known fraction, B, of the value $\epsilon_2 = \epsilon_{max}$ at the band maximum. The measured absorbances at λ_1, λ_2 and λ_3 are designated A_1, A_2 and A_3, respectively, and the difference between A'_2 (the true absorbance at λ_2) and A_2 is the background absorbance $(x + y)$. Thus

$$A'_2 = A_2 - (x + y)$$

where

$$x = (A_1 - A_3)(\lambda_3 - \lambda_2)/(\lambda_3 - \lambda_1)$$

and

$$y = (BA_3 - (A_2 - x))/(B - 1)$$

The wavelengths λ_1 and λ_3 must be chosen with some care. If $\lambda_3 - \lambda_1$ is large the assumption of a linear background over the interval is likely to be invalid. However, if λ_1 and λ_3 are very close, the absorbance values at these two wavelengths and at λ_2 will be rather similar, so that the differences between them may be comparable with the instrumental error. Morton and Stubbs developed their procedure specifically for the estimation of vitamin A in fish liver oils and, in these particular circumstances, chose λ_1 and λ_3 such that ϵ_1 and ϵ_3 were both 6/7 of the value of ϵ_2, giving a wavelength interval $\lambda_3 - \lambda_1$ of about 25 nm. The precision of the method, in the context of the estimation of vitamin A, has been examined in detail by Adamson *et al.* [3], who give estimates of the standard deviations likely to be encountered in careful spectrometric work.

The estimation of ergosterol, using the band at 282 nm, proves more difficult because the band is very sharp; a wavelength setting error of only 0.5 nm with λ_1 and λ_3 gives absorbance, and hence concentration, errors of about 20%. This led Shaw and Jefferies [4] to modify the Morton and Stubbs method, in that they selected λ_1 and λ_3 to coincide with adjacent maxima in the ergosterol spectrum. Although Morton and Stubbs [5] used the the particular condition $\epsilon_1 = \epsilon_3$, it is possible to apply the method in the more general case $\epsilon_1 \neq \epsilon_3$ as they and Shaw and Jefferies have pointed out. Although the relevant equations are more complex they are not difficult to use.

10.1.3 *Non-linear irrelevant absorption*

Not surprisingly, it is necessary to have methods available for dealing with non-linear irrelevant absorption, for those situations where there is non-linearity over a rather narrow wavelength range. Such non-linear backgrounds may be fitted to equations of the type

$$A = a + b\lambda + c\lambda^2 + d\lambda^3 + \dots$$

The characterization of such backgrounds requires absorbance measurements at not less than four wavelengths. However, the problem with these non-linear backgrounds is that they are often not of constant shape among a series of samples, as Ashton and Tootill [6] observed in the assay of griseofulvin. There is, therefore, the need for a different approach and this is provided by the use of harmonic analysis.

The basis of the method is that a given function, in this instance the absorption peak of the compound being estimated, can be expressed as the sum of a set of fundamental shapes. This is expressed by the equation:

$$f(\lambda) = ag_0 + bg_1 + cg_2 + dg_3 + \dots$$

where g_0 is constant, but g_1, g_2, g_3, etc., are functions of λ and a, b, c, d, etc., are coefficients. In harmonic analysis, g_1, g_2, g_3, etc., are trigonometric functions but Ashton and Tootill [7] and Glenn [8] have shown that Legendre polynomials are more convenient for computational purposes. These are also known as orthogonal polynomials. An absorption band is then expressed in the form

$$f(\lambda) = aP_0 + bP_1 + cP_2 + dP_3 + \dots$$

The first six Legendre polynomials are shown in Fig. 10.2 and it is evident that any peak may be represented as the sum of a constant background, a linear slope factor and a series of non-linear terms. If the irrelevant background absorption is expressed in a similar way, P_0 and P_1 terms will be present, together with some P_2 if the background is of quadratic form, but it is unlikely that the coefficient of P_3 will be appreciable. Hence, the coefficient in P_3 of the measured absorption curve is due solely to the component to be estimated and contains no contribution from the baseline. Furthermore it is directly proportional to its concentration and is used in place of the more usual ϵ. The procedure, therefore, is to express the measured absorption curve in terms of Legendre polynomials and to use one of these, probably P_3, for the determination. The simple computational details have been set out lucidly by Glenn [8]. The role of Legendre polynomials has been discussed in relation to wavelength calibration by Clewes [9].

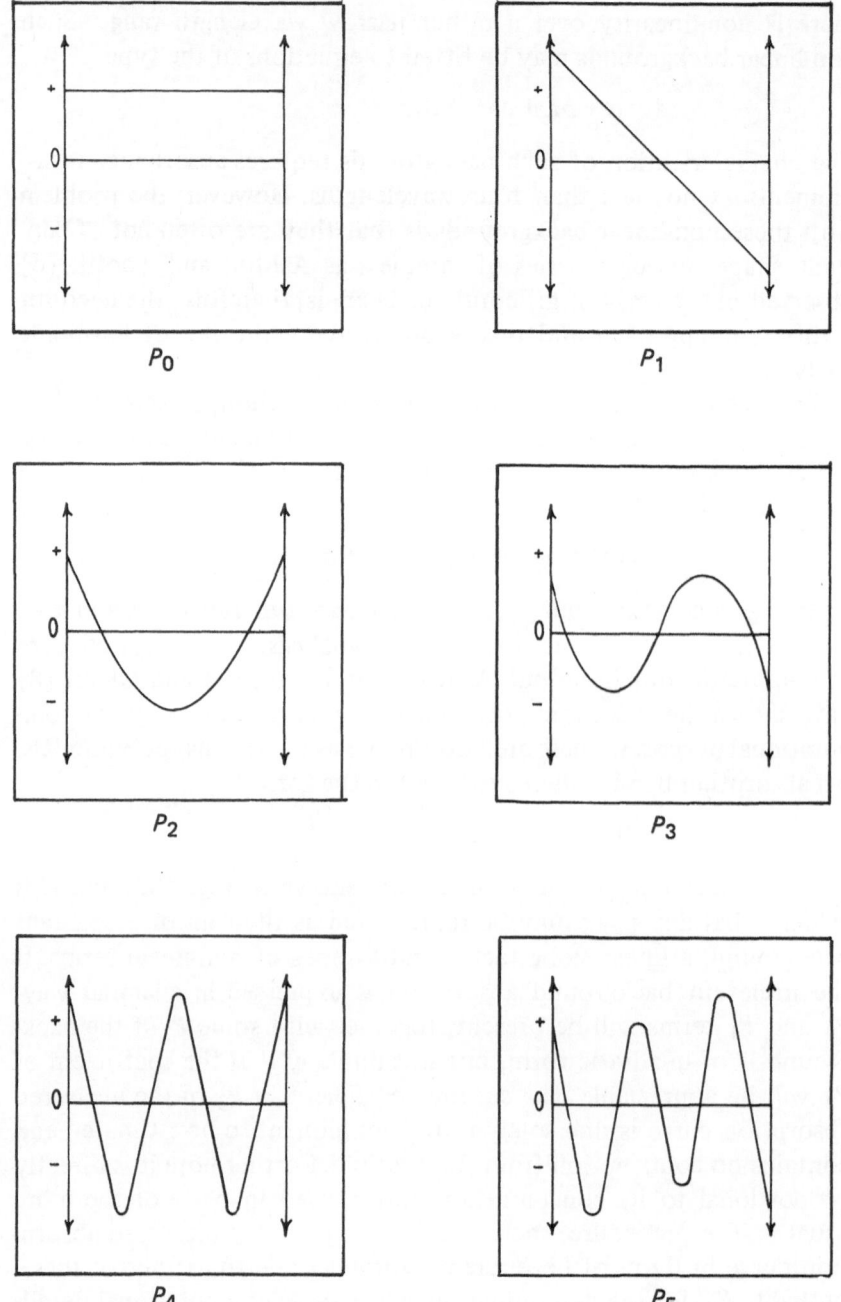

Fig. 10.2 *Representation of the first six Legendre polynomial functions. Any spectrum can be resolved into a sum of such functions. Redrawn from Clewes* [9].

10.2   Data smoothing

10.2.1 *Noise and smoothing*

Random fluctuations, usually referred to as noise, occur in all spectro-metric measurements. They arise largely, but not exclusively, in the detector system and associated electronics and are discussed in Chapter 6. This noise limits the precision that may be achieved in quantitative work and also sets the limit of detection of minor com-ponents. Electrical smoothing circuits are included in spectrometer detector and output systems but it is also advantageous to have the facility for mathematical smoothing. Not only does it provide greater flexibility, it operates on a fundamentally different basis to instru-mental smoothing. The latter can only process the information that has already passed through the system whereas mathematical smooth-ing on data collected in digital form provides the means for involving data points both before and after the one under consideration. Two methods are in general use for digital smoothing, the moving average and least squares methods.

10.2.2 *The moving average method*

Because noise is usually random in character, the averaging of a series of data points symmetrically placed about the one under considera-tion will lead to a reduction in the noise level. If the spectrometer output, e.g. absorbance, at three consecutive and equally separated wavelengths has the values A_1, A_2 and A_3, the simplest average is $(A_1 + A_2 + A_3)/3$. Alternatively, it may be useful to give more weight to the value of interest, A_2, by using an expression $(A_1 + 2A_2 + A_3)/4$. The smoothing operation consists of calculating such a three point average for λ_1, λ_2 and λ_3, then moving on to λ_2, λ_3, λ_4 until the last three points λ_{n-2}, λ_{n-1} and λ_n of the data set are reached. This is a very simple computational procedure and it is possible to use more than three points in the data set. However, this method assumes that there is linearity between A and λ over the interval used and if the wavelength interval is too large there will be an appreciable deviation and distortion.

Trott and Beynon [10] have recently extended the simple three-point moving average method, in two respects. They showed that if the process is repeated on the same data set there is a further improve-ment and, in principle, this should continue up to N passes, where $N = 0.5n - 1$ for n even and $0.5(n + 1) - 1$ for n odd, and n is the

number of points in the data set. However, they showed experiment-
ally that this smoothing efficiency limit is reached well before this
theoretical limit. They therefore further improved the method by
increasing the range covered by the three points. For example, if these
are denoted by A_{i-1}, A_i and A_{i+1} the spread can be increased to A_{i-2},
A_i and A_{i+2} and, more generally, to A_{i-j}, A_i and A_{i+j}. Clearly, the
greater the value of j the more effective will be the smoothing but
the greater will be the loss in resolution. This point has been examined
in detail by Maddams and Mead [11], whose results provide a clear
guide to the permissible degree of smoothing.

10.2.3 *The least squares method*

The effect of noise will be to lead to a scatter of individual data
points about the best estimated line drawn through them. Hence,
the determination of such a line, by a least squares procedure, provides an effective method for the reduction of random noise and this
is the basis of the widely used method of Savitzky and Golay [12]. A
set of n data points is fitted to a polynomial by a least squares calculation. The value of n is usually considerably smaller than the total
number of data points involved in defining the absorption band;
hence, this set of points is moved forward sequentially by one data
point until the whole band has been covered. This smoothing method
is mathematically equivalent to convoluting the original data with a
numerical function. Savitzky and Golay have provided values for these
functions for polynomials of various degrees and these make the
computer programming a straight forward task. Steinier *et al.* [13]
have subsequently noted errors in some of these numerical values and
have supplied corrected tables.

10.3   **Multicomponent analysis**

10.3.1 *Principles*

The absorbance of a solute at a particular wavelength, λ, is related to
its concentration and the pathlength by Beer's Law:

$$A_{1\lambda} = \epsilon_{1\lambda} bc$$

where ϵ_λ is the molar absorptivity at λ. If a second absorbing species
is present its absorption behaviour will be given by $A_{2\lambda} = \epsilon_{2\lambda} bc$ and,
if there is no interaction between the two components, the total
measured absorbance A_λ will be the sum of that of the two absorb-

ances, $A_{1\lambda} + A_{2\lambda}$. More generally, for an n component system, there will be the relation:

$$A_\lambda = \epsilon_{1\lambda} bc_1 + \epsilon_{2\lambda} bc_2 + \ldots + \epsilon_{n\lambda} bc_n.$$

Because there are n unknowns and only one equation it is not possible to solve for $c_1, c_2, c_3 \ldots c_n$. However, if measurements are made at n wavelengths the following set of simultaneous equations is obtained:

$$A_1 = \epsilon_{11} bc_1 + \epsilon_{21} bc_2 + \epsilon_{31} bc_3 + \ldots + \epsilon_{n1} bc_n$$

$$A_2 = \epsilon_{12} bc_1 + \epsilon_{22} bc_2 + \epsilon_{32} bc_3 + \ldots + \epsilon_{n2} bc_n$$

$$A_3 = \epsilon_{13} bc_1 + \epsilon_{23} bc_2 + \epsilon_{33} bc_3 + \ldots + \epsilon_{n3} bc_n$$

$$\ldots\ldots\ldots\ldots\ldots\ldots\ldots\ldots\ldots\ldots\ldots\ldots\ldots$$

$$A_n = \epsilon_{1n} bc_1 + \epsilon_{2n} bc_2 + \epsilon_{3n} bc_3 + \ldots + \epsilon_{nn} bc_n.$$

In principle, therefore, $c_1, c_2, c_3 \ldots c_n$ may be determined. The degree to which this may be achieved in practice will now be considered.

10.3.2 *The scope and limitations*

One potential limitation is readily appreciated by considering the simplest multicomponent system, a two component mixture, for which the two measured absorbances, A_1 and A_2, are given by the equations

$$A_1 = \epsilon_{11} bc_1 + \epsilon_{21} bc_2$$

$$A_2 = \epsilon_{12} bc_1 + \epsilon_{22} bc_2$$

Consider two extreme cases, the first in which component 2 does not absorb at λ_1 and component 1 does not absorb at λ_2. This gives the simple situation in which $A_1 = \epsilon_{11} bc_1$ and $A_2 = \epsilon_2 bc_2$ and the two components are, effectively, determined independently. The second extreme is when $\epsilon_{11} = \epsilon_{21}$ and $\epsilon_{12} = \epsilon_{22}$. There is then a total lack of specificity and only $c_1 + c_2$ may be determined. In choosing λ_1 and λ_2 it is therefore necessary to maximize the differences between the spectra of the two components, although it is unlikely that the ideal situation of zero absorption by one component will be achieved. It clearly becomes more difficult to maximize the specificity as the number of components increases and this factor, for a given error in the measurement of the various A and ϵ values, limits the precision that may be achieved.

It is common practice to choose λ_n, the characteristic wavelength for component n, in such a way that it lies at the maximum of a strong absorption band and is therefore insensitive to small errors in setting the monochromator. However, the way in which ϵ_{2n}, ϵ_{3n}, etc., vary with λ in the vicinity of λ_n must also be taken into account. Although these values may be appreciably smaller than ϵ_{1n}, if they vary rapidly with wavelength, because λ is located on the side of the absorption bands of other components, the fractional errors involved in their measurement may be considerably greater than for ϵ_{1n}.

These points are well illustrated in the analysis of four component mixtures of ethylbenzene and o-, m- and p-xylene (Fig. 10.3). The choice of the characteristic (or key wavelength as it is sometimes called) is, in principle, easiest in the case of p-xylene. This has its longest wavelength band at 274.7 nm a wavelength greater than those of the others, 271.0 nm for o-xylene and 272.7 nm for m-xylene as may be seen from Fig. 10.3. At 274.7 nm the absorptivities of ethylbenzene and o-xylene have fallen off considerably from the values at their maxima and they are changing comparatively slowly with wavelength. However, 274.7 nm is on the side of the m-xylene band at 272.7 nm and ϵ for the latter is changing rather rapidly with λ. Nevertheless, this must be tolerated as 274.7 nm is the best overall choice for p-xylene.

Likewise, 272.7 nm proves to be the best choice for m-xylene.

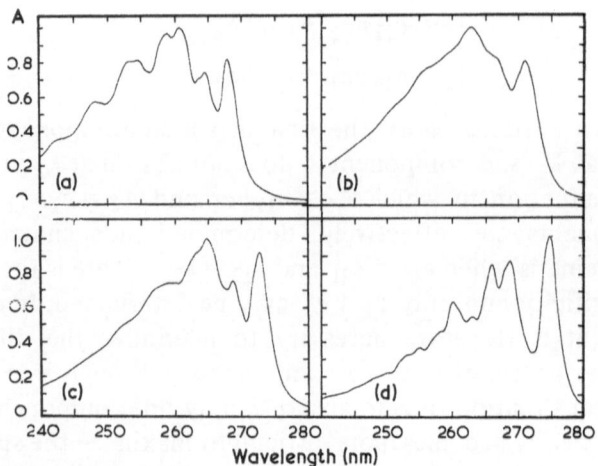

Fig. 10.3 *Absorption spectra of four related compounds. (a) Ethylbenzene, 0.5129 g l^{-1}; (b) o-xylene, 0.3925 g l^{-1}; (c) m-xylene, 0.3947 g l^{-1}; (d) p-xylene, 0.1707 g l^{-1}.*

Although ϵ is changing more rapidly with λ for ethylbenzene and *o*-xylene than at 274.7 nm because 272.7 nm is closer to their respective maxima, fortuitously this wavelength coincided with an absorption minimum in the *p*-xylene spectrum. 271.0 nm proves to be the best overall choice for *o*-xylene, although it is clear that the specificity will be lower for the material than for the *m*- and *p*-isomers, and because ϵ_{max} is smaller. Ethylbenzene proves to be the most difficult, because of the overall weakness of its spectrum by comparison with those of the xylene isomers. There is little choice other than to use one of the two most obvious bands, at 261.5 and 268.6 nm. The results of measurements on synthetic mixtures bear out what may be surmised from the spectra of the four components, that the precision of the *p*-xylene determinations is the highest and that of the ethylbenzene the lowest.

It will be evident from this example that the feasibility of a particular multicomponent analysis is dependent both on the number of components and the specificity of their spectra. It is possible to form a good idea of what may be achieved in practice by calculations based on propagation of errors, but with any system of this type it is highly advisable to undertake background studies with synthetic mixtures in order to establish the errors in the estimation of the various components. In practice, the errors may be larger because of the presence of irrelevant absorption at a low level or a minor well-defined component. For example, *iso*propylbenzene may be present in wider boiling xylene cuts and will interfere badly with the ethylbenzene determination.

Many three-component analyses have been reported and a few involving four components, but beyond this point other analytical methods, particularly based on separation procedures, become markedly superior.

When a particular multicomponent analysis has been shown to be feasible and is in frequent use, it is possible to simplify the calculations involving the use of simultaneous equations by making use of the properties of matrices. The set of equations considered above may be written in the matrix notation. For a given pathlength $b = 1$, the equation becomes $\mathbf{A} = E\mathbf{C}$, where the vector $\mathbf{A}$ contains all the absorbances A_1 to A_n and likewise the concentration vector, $\mathbf{C}$, the concentrations of the components c_1 to c_n. The matrix of molar absorptivities $\epsilon_n \lambda_n$ is denoted by E.

If this equation is premultipled by E^{-1} it gives $\mathbf{C} = E^{-1}\mathbf{A}$, so that the concentration values may be obtained directly. The inverse of this later matrix E^{-1}, is easily calculated by standard methods of matrix algebra, as discussed in various standards texts, e.g. Aitken [14] and Bauman [15].

The analysis of an n component mixture by inverting an absorbance matrix of n rows and columns is mathematically equivalent to fitting the measured spectrum at n points, the characteristic wavelengths, with the spectra of the components. It is possible to improve the precision of the analysis considerably by fitting the spectrum as a whole using a least squares criterion. In practice, this involves collecting absorbance values at close wavelength intervals over the whole of the spectral range of interest, preferably by digital means, and using a computer to obtain the least squares fit. Surprisingly, although this method was described almost 20 years ago [16] its merits have not been appreciated, but it is safe to predict that it will gain acceptance now that spectrometers with the requisite on-line data handling and computing facilities are coming into more general use.

There may be situations where deviations from the Beer's Law occur, for two reasons. Absorbance may not be a linear function of c, usually because of instrumental effects, such as the use of a slit-width which is not small by comparison with the bandwidth or because of the use of very high absorbance values. In addition there may be physical or chemical interactions between two or more components present in a mixture. Such interactions are more probable in IR spectroscopy, where concentrations of 10% are common place. Only in exceptional circumstances, such as highly hydrogen-bonding solvents and polar solutes, e.g. phenol in the presence of very low concentrations of hexamethyl phosphoramide, examined by Gerrard and Maddams [17], are interactions likely to occur at the low solute concentrations used in UV–VIS spectroscopy. When non-linearity occurs because of instrumental factors it is possible to obtain reasonably quantitative analysis by the use of approximation procedures; a very useful account has been given by Bauman [18] and the calculations involved are trivial when handled by a computer.

10.4   Matrix rank analysis

The basis of the use of simultaneous equations for the analysis of multicomponent systems is that the number of components and their identities are known, but with unknown systems neither are available. Although analysis by other techniques will often provide the information it may be possible to identify the components by UV–VIS spectroscopy by, using 'spectral stripping' or curve fitting techniques (see Section 10.5). Matrix rank analysis or 'factor analysis' can often provide this information more rapidly.

If absorbance measurements are made at a number of wavelengths on a series of solutions containing an unknown number of components whose concentrations vary with respect to each other, the results may be set out in matrix form. For example, for i such solutions and measurements at j wavelengths, the ij measurements constitute the rows and columns, respectively, of the matrix A, where

$$A = \begin{bmatrix} A_{11} & A_{21} & A_{31} & \ldots & A_{n1} \\ A_{12} & A_{22} & A_{32} & \ldots & A_{n2} \\ A_{13} & A_{23} & A_{33} & \ldots & A_{n3} \\ \cdot & \cdot & \cdot & & \cdot \\ \cdot & \cdot & \cdot & & \cdot \\ \cdot & \cdot & \cdot & & \cdot \\ A_{1n} & A_{2n} & A_{3n} & \ldots & A_{nn} \end{bmatrix}$$

Such a matrix may be square, i.e. $i = j$, or rectangular, $i \neq j$. Each element of the matrix is the sum of the absorbances of the individual components in a particular solution at one wavelength. As noted in the previous section, the absorbance matrix is the product of two separate matrices, one of which C represents the concentrations of each component in the different mixtures and a second E which represents the molar absorptivities of each component at the various wavelengths.

If the number of wavelengths chosen and the number of solutions examined is greater than the anticipated number of components, at least one row or column of elements in the C and E matrices will consist wholly of zeros. However, this will not be apparent from the absorbance matrix, as is evident from the following simple example:

$$\begin{bmatrix} 7 & 17 & 26 \\ 10 & 14 & 24 \\ 13 & 11 & 22 \end{bmatrix} = \begin{bmatrix} 3 & 1 & 0 \\ 2 & 2 & 0 \\ 1 & 3 & 0 \end{bmatrix} \times \begin{bmatrix} 1 & 5 & 7 \\ 4 & 2 & 5 \\ 0 & 0 & 0 \end{bmatrix}$$

$$ A C E$$

The above example might well represent the results of measurements on a two component system, measuring at three wavelengths on three different solutions. The number of components present is thus given by the rank, R, of the A matrix, which is defined as the order of the largest non-zero determinant computed from the matrix components, while all other determinants of order $R + 1$ are zero. This is a direct consequence of the fact that if the matrix is of rank R it requires the existence of R linearly independent absorbing components.

The value of R may be found by testing all of the determinants in the matrix for zero values. This is a time-consuming process, although it is readily amenable to computer calculation, and a further complication arises because measurement errors mean that a zero value is very rarely obtained. It is then necessary to use statistical tests to determine which are zero values. The method of Wallace and Katz [19] involves the setting up of a second comparison matrix whose elements S_{ij} are the estimated errors in the observed values of A_{ij}. The A_{ij} matrix is then transformed to an equivalent matrix, whose elements below the principal diagonal are all zero. The error matrix S_{ij} is also transformed during the reduction of A_{ij}, by calculating new values for A_{ij} based on the propagation of errors during the transformation of A_{ij}. The rank of A_{ij} is then found from the number element being given by the transformed S_{ij} matrix. Worked examples, which illustrate the method very clearly, are given by Wallace and Katz, who used the method to determine the number of species present in methyl red solutions of different pH values, and by Gerrard and Maddams [20], who examined phenol in solution in cyclohexane containing various concentrations of ethanol.

Although it should be possible to determine the rank of a matrix and the number of components present in a system by using the number of wavelengths equal to the rank, it is advisable with a completely unknown system to use a number of wavelengths significantly in excess of the likely number of components, for two reasons. It ensures that a more than adequate number of absorbance values is available and, as some wavelengths are more likely to be useful than others in terms of specificity for the individual components, it also increases the reliability on this count. The fact that the matrix is large is of no consequence because the calculations involved are readily handled by a minicomputer. However, it must be emphasized that matrix rank analysis will not give a greater degree of discrimination than is inherent in the measurements. If two components have very similar spectra, or a component is present in low concentration, there will be ambiguity. Two useful papers, by Ainsworth [21] and Burgess

[22], deal with the application of matrix rank analysis to UV–VIS spectrometry, and Malinowski and Howery have recently published a detailed work on factor analysis [23].

10.5 Spectral stripping and related techniques

The very high performance of modern spectrometers particularly with respect to sensitivity and signal-to-noise ratio, coupled with the availability of computers, off- or on-line, has led to the increased use of other mathematical methods for extracting information from the spectra of mixtures of components.

Spectral stripping provides a convenient introduction to these newer approaches, both as a method and because it provides a link between the manual and computer techniques. If the spectrum of a multicomponent mixture gives one or more clearly defined bands that can be assigned to specific compounds, these bands can be sequentially subtracted from the spectrum, giving a quantitative estimation, and also simplifying the residual spectrum so that it may be further interpreted. The subtraction can be done instrumentally or by computer. In the former case a sample of the first of the identified components is placed in a variable pathlength cell in the reference beam of the spectrometer and the thickness is adjusted until the relevant band or bands in the mixture, sited in the sample beam, disappear. The process is then repeated with the second, and possibly other, components. The alternative approach of computer subtraction involves running the spectra of the identifiable components, storing them in digital form in the computer and using a scaling factor calculated by the computer to give exact subtraction from the composite spectrum. This approach offers greater flexibility and is less time consuming; it is a cogent example of the logical use of the computer in spectroscopy.

This technique is simply a multiple application of difference spectroscopy but has limitations, some obvious and some a trap for the unwary. It will clearly work to best advantage when the overall spectrum shows one or more well defined bands that are not badly overlapped. This is a necessary but not sufficient condition, because the species responsible for these bands must not only be identified but must be available. As with the use of simultaneous equations it is tacitly assumed that the Beer's Law is applicable. If there are interactions which lead to small shifts in the positions of the component bands in the mixture spectrum the subtraction of the bands of the individual components can never be exact. The results will be akin to the shape of a first derivative spectrum. Fortunately, this effect is usually recognizable. During the subtraction process the errors in both the original data and in the inexactness of the subtraction become cumulative, and it is particularly important to avoid working with bands of very high or low intensity, as the spectrometric errors in the original spectrum will be greater. Finally, it is unwise to try to push the technique beyond reasonable limits. These cannot be defined uniquely as they vary from situation to situation. Perhaps the best working rule is that if the spectrum after a subtraction looks unusual or distorted, an error should be suspected.

Spectral stripping relies on the sequential subtraction of a series of identifiable peaks from a composite spectrum. It is possible to work in the opposite direction and match a multipeaked spectrum by adding a series of computer-generated peaks until the resulting synthetic spectrum matches the experimental data. This process is known as curve fitting, and it is also subject to appreciable limitations, as has become increasingly recognized [24–27]. These limitations may be detailed by reference to Fig. 10.4 which shows a composite profile, together with the results obtained by curve fitting. This has been done in terms of three peaks, a reasonable approach in view of the visual appearance of the composite profile. However, it is pertinent to ask if four or more peaks would have been more appropriate and, therefore, what criteria exist for establishing the number of peaks. It is also desirable to have approximate values for their locations on the wavelength axis, information that is usually obtained by visual inspection.

It is also necessary to establish two other sets of parameters for peaks, either by assumption or by an optimization procedure, as part of the calculations. These are the band shapes and half-widths. In the present instance fitting has been done with peaks having the Lorentzian shape and of equal half-width but, in practice, it is necessary to justify a particular assumption. In Fig. 10.4 it has been necessary to include a baseline in addition to the three peaks, and this generally proves to be the case. The baseline has been taken as linear but this may not be warranted. Hence methods are required to define baselines and to deal with the general situation when they are not linear. Finally, a goodness of fit criterion is required to assess the overall success of the curve fitting and to show if any of the assumptions are seriously in error.

Most curve fitting studies have related to the characterization of

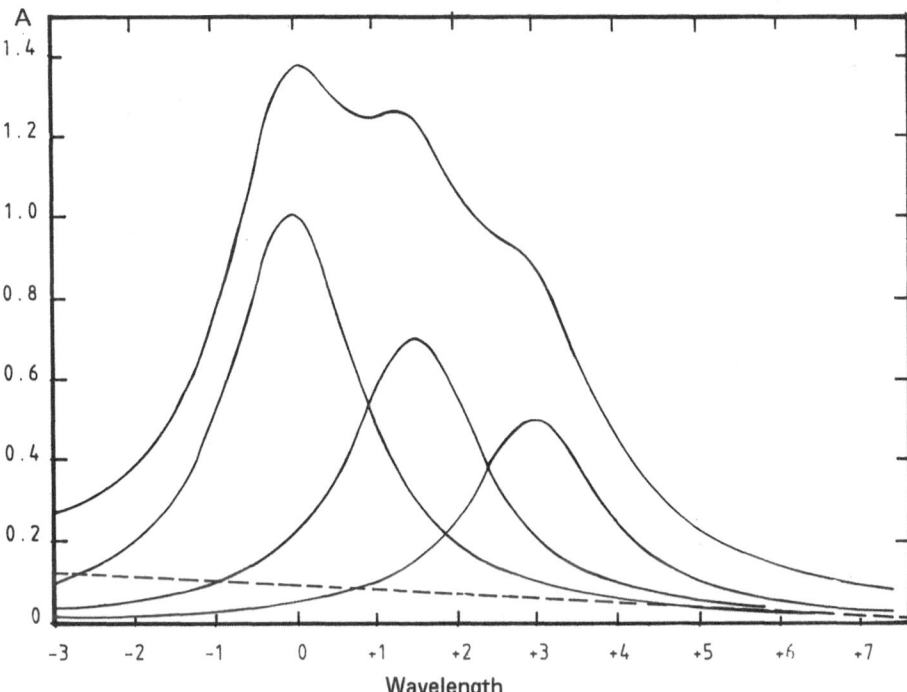

Fig. 10.4 *Fitting a complex spectral shape by summation of components bands. The upper curve has been fitted by the sum of the three simple curves plus a linear sloping background.*

overlapping vibrational bands in IR and Raman spectra, and there is a definite change of emphasis when dealing with overlapping UV−VIS absorption bands. The band shape is different, the half-widths are much greater and the bands are more overlapped but tend to be fewer in number. The primary spectral data have a markedly superior signal-to-noise ratio in the UV−VIS region. Some of these factors facilitate the use of curve fitting, some make it more difficult, and with others such as the band shape, the problem is primarily that of justifying assumptions. These factors have been discussed in detail by Barker and Fox [28], who have done pioneering work in this field.

The greater the number of parameters to be optimized in the least squares process of curve fitting, the lower will be the probability that a unique solution will be forthcoming. It is therefore advantageous to fix, or at least to constrain, some of these parameters, and the number of component peaks is the one that offers the best possibilities. Visual inspection will often reveal almost hidden peaks, that appear as slight shoulders, and the band sharpening achieved by the use of even-numbered derivatives is also valuable for estimating the number of component peaks. Recently, a promising new approach to band sharpening, termed Fourier self-deconvolution, has emerged. Any experimental peak can be expressed, mathematically, as a convolution of a lineshape function and a spectral function. If part or all of the former can be deconvoluted from the measured peak a considerable sharpening, by as much as a factor of three, results. Hence in the case of overlapped peaks resolution of the components may be achieved to a degree that is valuable for peak finding prior to curve fitting.

On grounds of mathematical convenience the deconvolution is undertaken in the Fourier domain and it is an added advantage that an increasing number of IR and NMR spectrometers have their primary output in this form. When the spectrometer output is in the conventional form of absorption intensity as a function of wavelength it is collected in digital form and transformed by the use of the equation:

$$A(\lambda) = \int_{-\infty}^{\infty} I(x) \exp (i2\pi\lambda x) \, dx = F\{I(x)\}$$

The last parameter, $F\{I(x)\}$ is the Fourier transform of $A(\lambda)$. Such calculations present no problems with modern computers. The paper of Kauppinen *et al.* [29] should be consulted for fuller details. As yet, the technique has been used only with systems of overlapping vibrational bands but it could prove useful in the case of composite band systems in UV–VIS spectra.

References

1 Mulder, E.J., Spruit, F.J. and Keuning, K.J. (1963), *Pharm. Weekblad.*, **98**, 745.
2 Morton, R.A. and Stubbs, A.L. (1947), *Biochem. J.*, **41**, 525.
3 Adamson, D.C.M., Elvidge, W.F. Gridgman, N.T., Hopkins, E.H., Stuckey, R.E. and Taylor, R.J. (1951), *Analyst*, **76**, 445.
4 Shaw, H.C. and Jefferies, J.P. (1953), *Analyst*, **78**, 519.
5 Morton, R.A. and Stubbs, A.L. (1946), *Analyst*, **71**, 348.
6 Ashton, G.C. and Tootill, J.P.R. (1956), *Analyst*, **81**, 232.
7 Ashton, G.C. and Tootill, J.P.R. (1956), *Analyst*, **81**, 225.
8 Glenn, A.L. (1963), *J. Pharm. Pharmacol. Suppl.*, **15**, 123T.
9 Clewes, B.N. (1979), *UV Spectrom. Grp Bull.*, **7**, 35.
10 Trott, G.W. and Beynon, J.H. (1979), *Int. J. Mass. Spectrom. Ion Phys.*, **31**, 37.

11 Maddams, W.F. and Mead, W.L. (1982), *Spectrochim. Acta*, **38A**, 437.

12 Savitzky, A. and Golay, M.J.E. (1964), *Anal. Chem.*, **36**, 1627.

13 Steinier, J., Termonia, Y. and Deltour, J. (1972), *Anal. Chem.*, **44**, 1906.

14 Aitken, A.C. (1944), *Determinants and Matrices*, 3rd Edn, Oliver and Boyd, Edinburgh and London.

15 Bauman, R.P. (1962), *Absorption Spectroscopy*, Appendix 1, *Matrix Methods*, Wiley, New York.

16 Blackburn, J.A. (1965), *Anal. Chem.*, **37**, 1000.

17 Gerrard, D.L. and Maddams, W.F. (1978), *Spectrochim. Acta*, **34A**, 1219.

18 Bauman, R.P. (1962), *Absorption Spectroscopy*, Wiley, New York, 413–19.

19 Wallace, R.M. and Katz, S.M. (1964), *J. Phys. Chem.*, **68**, 3890.

20 Gerrard, D.L. and Maddams, W.F. (1978), *Spectrochim. Acta*, **34A**, 1213.

21 Ainsworth, S. (1972), *Photoelec. Spectrom. Grp Bull.*, **20**, 611.

22 Burgess, C. (1979), *UV Spectrom. Grp Bull.*, **7**, 25.

23 Malinowski, E. R. and Howery, D.G. (1980), *Factor Analysis in Chemistry*, Wiley, New York.

24 Vandeginste, B.G.M. and DeGalan, L. (1975), *Anal. Chem.*, **47**, 2124.

25 Morrow, B.A. and Cody, I.A. (1973), *J. Phys. Chem.*, **77**, 1465.

26 Gans, P. and Gill, J.B. (1977), *Appl. Spectrosc.*, **31**, 451.

27 Maddams, W.F. (1980), *Appl. Spectrosc.*, **34**, 245.

28 Barker, B.E. and Fox, M.F. (1980), *Chem. Soc. Rev.*, **9**, 143.

29 Kauppinen, J.K., Moffatt, D.J., Mantsch, H.H. and Cameron, D.G. (1981), *Appl. Spectrosc.*, **35**, 271.

11 Special techniques

The generally uninformative nature of the electronic envelope, coupled with the broad, overlapping nature of constituent absorption bands in mixtures, has led to the search for technical modifications, or spectroscopic 'tricks', for improving specificity and accuracy. Perhaps the earliest successful addition to the repertoire was Britton Chance's technique of dual-wavelength spectrometry [1] for measurements of turbid samples in biochemical analysis, later to be extensively applied by Shozo Shibata in analytical spectroscopy [2]. Shortly after Chance's pioneering work, two groups independently proposed the use of mathematical derivative functions in spectroscopy. In America, Giese and French [3] developed the earliest first-derivative device for examining the visible spectra of pigments in plant photochemistry. At the same time in the UK, Collier and Singleton patented the idea of using second- and higher-order derivative spectra [4], illustrated by their work in IR spectroscopy [5]. Difference spectroscopy, by contrast, has long been used for spectral correction purposes, represented in its simplest form by subtractive compensation of sample absorbance with that of the solvent. Recent improvements in instrumentation have led to the use of difference spectroscopy in the measurement of highly absorbing samples.

The impact of low-noise operational amplifiers and of microcomputers on the practice of UV–VIS spectroscopy has been profound (cf. Chapter 7), not only as regards instrument control, but also with respect to the acquisition, storage and manipulation of digitized spectral data. A number of the spectroscopic 'tricks' discussed below have been made more accessible by virtue of digital techniques, which have undoubtedly transformed the practice of UV-visible spectroscopy.

11.1 Derivative spectroscopy

The derivative method is one of several generally applicable methods available for the transformation of spectral data. In this method the transmittance, or more appropriately, the absorbance (A) of a sample in solution or in the gaseous state is differentiated with respect to wavelength, to generate the first-, second- or higher-order derivative:

$$A = f(\lambda)$$

$$dA/d\lambda = f'(\lambda)$$

$$d^2 A/d\lambda^2 = f''(\lambda) \ldots$$

Spectra transformed in this way often yield a more characteristic profile, where subtle changes of gradient and curvature are observed as distinctive bipolar features.

The first derivative of a UV−VIS absorption band, which represents the gradient at all points of the spectral envelope, has often been used to detect and locate 'hidden' peaks (since $dA/d\lambda = 0$ at peak maxima) and as a characteristic profile for identification [3, 6]. However, it turns out that the second derivative, which describes the 'curvature' of the original band, and the higher even derivatives are potentially more useful analytically [7].

The odd derivatives of a Gaussian band are observed as bipolar disperse functions dissimilar to the original band (Fig. 11.1). The even derivatives, however, are seen as bipolar functions of alternating sign at the centroid, whose position coincides with that of the original band maximum. To this extent even derivative spectra bear a similarity to the original spectrum, although the presence of satellite peaks flanking the centroid adds a degree of complexity to the derivative profile. A key feature of derivative spectroscopy is the observation that, relative to the original spectral bandwidth in the so-called 'zero-order spectrum', the derivative centroid bandwidth decreases to 53%, 41% and 34% in the second, fourth and sixth orders, respectively [7]. This feature of the derivative process forms the basis of resolution for overlapping bands [8, 9], although it is clear that the band-sharpening process in Gaussian bands is less marked in the fourth and higher orders. Moreover, the increasingly complex satellite patterns detract from resolution enhancement in the higher derivative spectra.

An important property of the derivative process is that broad bands are suppressed relative to sharp bands, to an extent that increases with derivative order. This arises from the observation that the

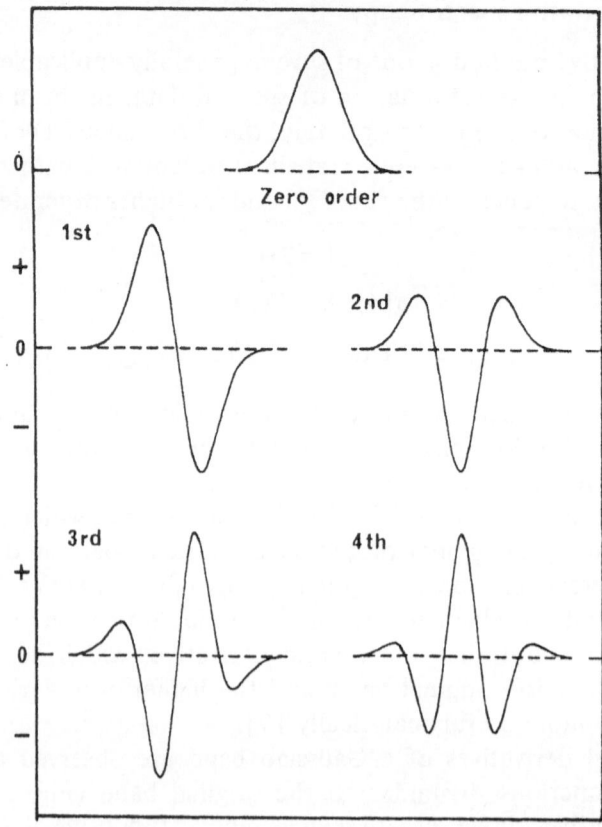

Fig. 11.1 *A Gaussian absorption band and its first to fourth order derivatives.*

amplitude D_n of a Gaussian (or Lorentzian) band in the nth derivative is inversely related to the original bandwidth W, raised to the nth degree:

$$D_n \propto W^{-n}$$

Thus for two concident bands of equal intensity, the nth derivative amplitude of the sharper band X is greater than that of the broader band Y by a factor which increases with derivative order:

$$\frac{D_{n,\,X}}{D_{n,\,Y}} = \left(\frac{W_Y}{W_X}\right)^n$$

This property leads to the selective rejection of broad, additive spectral interferences and can lead to increased detection sensitivity in the second- and fourth-order derivatives [7,9–11]. Other types of background interference, such as 'irrelevant absorption' or Rayleigh scattering, are also selectivity suppressed in the derivative spectrum (Fig. 11.2).

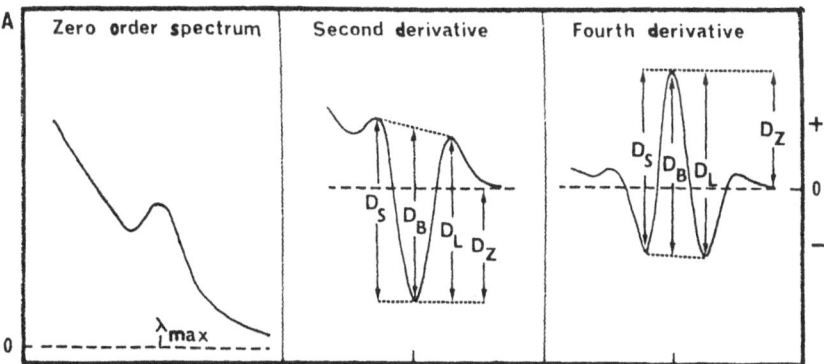

Fig. 11.2 *A Gaussian absorption band superimposed on a light-scattering matrix* $(A \propto \lambda^{-4})$, *and its second and fourth derivatives, illustrating the graphical measures available for quantitative assay.* D_S *and* D_L *are the amplitudes of the short- and long-wavelength satellites,* D_B *is the mid-point amplitude and* D_Z *is the derivative zero. These values are related to concentration through the derivative form of Beer's Law.*

The quantitative relationships which apply in derivative spectroscopy depend on whether transmittance or absorbance is the function of radiation intensity recorded. If transmittance is employed, the resulting expressions after differentiation become non-linear with concentration in the second and higher derivatives [7, 10]. If, however, absorbance (A) is employed, and if Beer's Law is obeyed at any defined wavelength λ in the zero-order spectrum:

$$A = \epsilon bc \qquad \text{at } \lambda$$

then

$$\frac{dA}{d\lambda} = \frac{d\epsilon}{d\lambda} bc \qquad \text{at } \lambda$$

and

$$\frac{d^n A}{d\lambda^n} = \frac{d^n \epsilon}{d\lambda^n} bc \qquad \text{at } \lambda$$

where ϵ is the molar absorptivity ($\text{M}^{-1} \text{cm}^{-1}$), b is the cell pathlength (cm) and c is the concentration (mol l^{-1}).

For quantitative work the amplitude of a derivative peak can be measured graphically in various ways as illustrated in Fig. 11.2 [7, 12]. Although the 'true' derivative amplitude is that measured with respect to the derivative zero, the most common practice is to record the amplitude with respect to satellite features of the spectrum (Fig. 11.2). This method affords an extra degree of suppression of matrix interference, since it is effectively a form of internal normalization of the derivative peak with respect to the overall derivative spectral envelope [7]. For highest accuracy it is common practice to run standards in bracketting sequence with samples, subjecting both to the same experimental conditions. It should have been established that the graphical derivative measure adopted fulfils the conventional analytical criteria of linear response with concentration, regression through or close to the origin, independence from matrix or other interferences and optimum precision.

11.1.1   *The production of derivative spectra*

In general, methods for generating derivative spectra fall into two classes: optical methods which operate on the radiation beam itself; and electronic or digital methods operating on the photometric detector output. The principal optical method is represented by the wavelength modulation technique, where the wavelength of incident radiation is rapidly modulated over a narrow wavelength range $\Delta\lambda$ by an electromechanical device (oscillating slits, grating or mirror). This method has found popularity in the construction of dedicated spectrometers for environmental monitoring, where the characteristic second derivative profile of polynuclear aromatic hydrocarbons, for example, aids rapid identification [13]. In another optical technique, the dual-wavelength spectrometer has been used to generate first derivative spectra, by scanning the spectrum with each monochromator separated by a small, constant interval $\Delta\lambda$, as discussed below [2].

The second and higher derivatives can, however, be more readily generated using low-noise analogue resistance–capacitance (RC) devices [10] or by digital techniques [12, 14, 15]. The electronic analogue device generates the required derivative as a function of time as the spectrum is scanned at constant speed $(d\lambda/dt)$:

$$\frac{dA}{dt} = \frac{dA}{d\lambda}\frac{d\lambda}{dt}$$

$$\frac{d^2A}{dt^2} = \frac{d^2A}{d\lambda^2}\left[\frac{d\lambda}{dt}\right]^2$$

The 'true' value of the derivative is therefore related to the derivative amplitude observed under defined instrumental conditions, through an instrumental constant, K:

$$\frac{\mathrm{d}^n A}{\mathrm{d}\lambda^n} = K \frac{\mathrm{d}^n A}{\mathrm{d}t^n}$$

The analogue derivative amplitude is observed to vary with the following instrumental parameters: scan speed, slitwidth, RC device gain factor. Furthermore, the signal-to-noise ratio is found to degrade by approximately a factor of two in each successive derivative order [12], so that the conventional RC device is limited to generating fourth derivatives [7, 11]. This limit has been extended to eighth and ninth orders by the use of special electronic circuitry for enhanced performance in the higher derivatives [10].

An alternative method for generating derivative spectra is based on the microcomputer, employing one of a number of digital algorithms [9, 12, 14, 15] to produce smoothed spectral derivatives, either in real time or by post-run processing of the digitized spectrum. The digital approach is increasingly employed in contemporary spectrometers, due to the widespread adoption of microprocessors for instrument control and data handling, coupled with the addition of dedicated microcomputers for further data processing capability.

It should be noted that although transformation of a UV–VIS spectrum to its second or higher derivative often yields a more highly characteristic profile than the zero-order spectrum, the intrinsic information content of the data is not increased – indeed, some data, such as constant 'offset' factors, are lost. Rather does the derivative method tend to emphasize subtle spectral features barely detectable by eye, by presenting them in a new and visually more accessible way (Fig. 11.3). Moreover, the derivative method is generally applicable in analytical chemistry and can be used equally for resolution enhancement of chromatographic [7] or densitometric [16] data.

Derivative UV–VIS spectroscopy has found significant application in the environmental [13], pharmaceutical [10, 17–20], clinical [21], forensic [22], biomedical [10, 23] and industrial areas [11]. The method can be combined with difference or with dual-wavelength spectroscopy, to give enhanced discrimination against matrix interference, as discussed below.

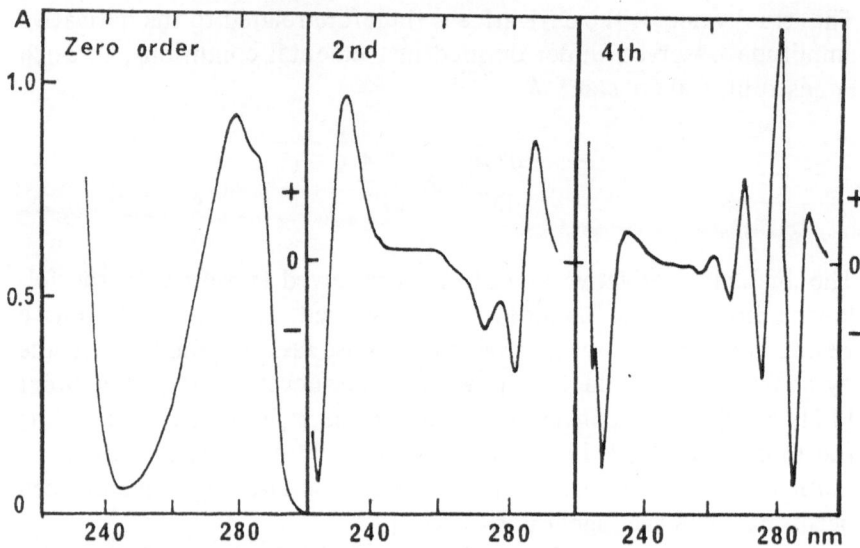

Fig. 11.3 *The absorption spectrum (zero order) and second and fourth derivatives of ethinyl oestradiol in methanol (140 μg ml^{-1}).*

11.2 Difference spectroscopy

In conventional spectrometry, the absorbance of a liquid sample in a suitable cell is compensated by automatic (or manual) subtraction of solvent (A_s) and window (A_w) absorption at a defined wavelength, using a matched cell filled with the appropriate solvent. In this sense, all absorbance measurements in UV−VIS spectrometry are made by means of difference spectroscopy, where the difference in absorbance, ΔA, is equivalent to the absorbance of the analyte alone, A_X :

$$\Delta A = A_X + A_s + A_w - (A_s + A_w) = A_X \qquad \text{at } \lambda.$$

The difference spectroscopic method is sometimes described as 'differential spectroscopy', an ambiguous term not to be recommended in view of the increasing exploitation of derivative spectroscopy (cf. Section 11.1).

In general, difference spectroscopy involves the measurement of an absorbance difference between a liquid sample and a reference solution. The latter may consist of the sample solution in physically or chemically changed form, or it may comprise a solution equivalent in composition to the liquid matrix in which the sample is located.

Two of the major types of application are represented by (a) matrix compensation methods, and (b) high-precision measurement of high absorbance values. For convenience, each application type will be considered separately.

11.2.1 *Matrix compensation methods*

These methods generally exploit a change in the chemical or physical properties of the analyte X, which permits its selective detection in the presence of interfering matrix M. The assumption is made that the spectrum of the analyte can be changed without affecting the matrix spectrum. If, for example, the analyte is pH-sensitive at a defined wavelength, with molar absorptivity values of ϵ_X at pH_1 and ϵ'_X at pH_2, and if the effective molar absorptivity of the matrix, ϵ_M, remains constant at both values of pH, then assuming that the law of additivity of absorbances applies, the total absorbance A_T observed at pH_1 and A'_T observed at pH_2 becomes:

$$A_T = \epsilon_X bc_X + \epsilon_M bc_M$$

$$A'_T = \epsilon'_X bc_X + \epsilon_M bc_M$$

The difference absorbance, ΔA (measured with the less absorbing solution in the reference cell), becomes:

$$\Delta A = |A_T - A'_T| = |\epsilon_X - \epsilon'_X| bc_X \quad \text{and} \quad \Delta A = \Delta\epsilon_X bc_X$$

Values for $\Delta\epsilon_X$ can be established by prior standardization, or *in situ* for every assay. If Beer's Law holds at the analytical wavelength employed, then the concentration of analyte in a test can be readily found by simple proportion:

$$\Delta A_{\text{test}} / \Delta A_{\text{standard}} = C_{X(\text{test})} / C_{X(\text{standard})}$$

Many suitable methods for physical and chemical modification of the analyte absorbance have been reported. As an example, the classical method described by Görög for the determination of Δ^4- and $\Delta^{1,4}$-3-ketosteroids in pharmaceutical applications, illustrates the general approach involved [24]. A methanolic solution of a sample containing prednisolone is reduced by sodium borohydride to yield the spectrally inactive 3-hydroxy derivative. The residual absorption (A'_P) at about 240 nm is equivalent to the absorption of the unchanged matrix (A_M), so that when the reduced solution is placed in the reference cell the untreated prednisolone sample absorbance (A_P) is compensated by difference for the matrix interference:

$$A_T = A_P + A_M$$

$$A'_T = A'_P + A_M$$

$$\Delta A = A_T - A'_T = A_P \qquad \text{since } A'_P = 0$$

This elegant procedure has been modified by Chafetz *et al.* [25], who substituted lithium borohydride in tetrahydrofuran for more efficient reduction of $\Delta^{1,4}$-3-ketosteroids. In order to establish that the matrix spectrum is unchanged by the reducing agent, a plot of log (ΔA) as a function of wavelength can be compared with the logarithmic spectrum of the pure component, these curves being superimposable if the matrix interference has been eliminated. This follows from the observation that the shape of a logarithmic spectrum is defined only by the molar absorptivity as a function of wavelength, the concentration term being expressed as a vertical displacement along the log A axis:

$$\log_{10}(\Delta A) = \log_{10}(\Delta\epsilon) + \log_{10}b + \log_{10}c \qquad \text{at } \lambda$$

Indeed, small differences between the two curves could be accentuated by transforming each log (ΔA) spectrum to its second derivative [7].

In analytical biochemistry, where difference spectroscopy is widely used for investigations on the effect of solvent, heat and other perturbing factors, transformation of the difference spectrum to its first or second derivative can be successfully achieved, and has permitted the examination of tyrosyl residues resolved in cytochrome P450 [26] and phenylalanine residues resolved in protein spectra, respectively [27]. Second derivative-difference spectroscopy, as the technique is described, affords a further method for rejecting matrix interference, while simultaneously sharpening any fine structural features and improving their resolution from adjacent bands.

In its simplest form, difference spectroscopy is practised in industrial quality control, in those cases where the sample matrix is well-defined. An appropriate dilution of the matrix is placed in the reference cell, on the assumption that the matrix composition can be accurately replicated and is not subject to variation during the industrial process. The difference absorbance is, however, susceptible to systematic error introduced by any uncertainty in the concentration of the matrix in the sample to be assayed. This error increases in proportion to the ratio of the molar absorptivities of matrix to analyte.

11.2.2 *High precision measurements*

High precision measurements of highly absorbing solutions can be achieved by using as reference a standard solution slightly lower in concentration than the test sample, coupled with expansion of the transmittance or absorbance scale and a wider slitwidth to increase energy throughput [28]. In principle, if the analyte obeys Beer's Law up to the absorbance value of test (A_t) and standard (A_s) then the difference absorbance, ΔA, is related to analyte concentration in test (C_t) and standard (C_s) as follows:

$$\Delta A = A_t - A_s = \epsilon b(C_t - C_s) = \epsilon b \Delta c.$$

In practice, stray-light, instrument noise and limits of photometric linearity require that a calibration curve of ΔA versus Δc be established. Various modifications of this technique have been proposed for high-sensitivity trace analysis [29].

11.3 Dual-wavelength spectroscopy

In this method, two independently monochromated beams of radiation at λ_1 and λ_2 are time-shared through a single sample cell. The difference in absorbance, ΔA, is recorded at the two analytical wavelengths selected, with the aim of reducing or eliminating matrix interference or for the determination of analytes in a multicomponent mixture:

$$\Delta A = A^{\lambda_1} - A^{\lambda_2}$$

The sample cell is usually positioned close to the detector so that turbidity or scattered radiation can be effectively compensated [30].

Dual-wavelength spectroscopy has been shown to permit accurate measurement of small absorbance differences, both in highly absorptive solutions and in very weakly absorbing systems [2, 30]. The 'reference' wavelength, λ_2, confers great flexibility on the method. In the presence of Rayleigh scattering, for example, the reference wavelength is set to a point on the spectral profile, which is equi-absorptive with the anticipated scattering contribution, $A_R^{\lambda_1}$, at the 'analytical' wavelength, λ_1:

$$A_T^{\lambda_1} = A_X^{\lambda_1} + A_R^{\lambda_1} \qquad \text{at } \lambda_1$$

$$A_T^{\lambda_2} = A_R^{\lambda_1}$$

$$\Delta A = A_T^{\lambda_1} - A_T^{\lambda_2} = A_X^{\lambda_1}$$

where the subscripts T, X and R refer to the total absorbance, and the analyte and Rayleigh scattering contributions, respectively (Fig. 11.4).

A further example of matrix correction is the case where the reference wavelength is set at a point λ_2 in the spectrum where the matrix absorption, $A_M^{\lambda_1}$, behaves similarly to the interference, $A_M^{\lambda_2}$, observed at the analytical wavelength, λ_1:

$$A_T^{\lambda_1} = A_X^{\lambda_1} + A_M^{\lambda_1}$$
$$A_M^{\lambda_2} = A_M^{\lambda_1}$$
$$\Delta A = A_T^{\lambda_1} - A_M^{\lambda_2} = A_X^{\lambda_1}$$

A variant of this method involves scanning the spectrum in difference mode, keeping the reference wavelength constant at a spectral position which constitutes a reference for the system [2]. Such 'relative' spectra correspond to a form of internal normalization.

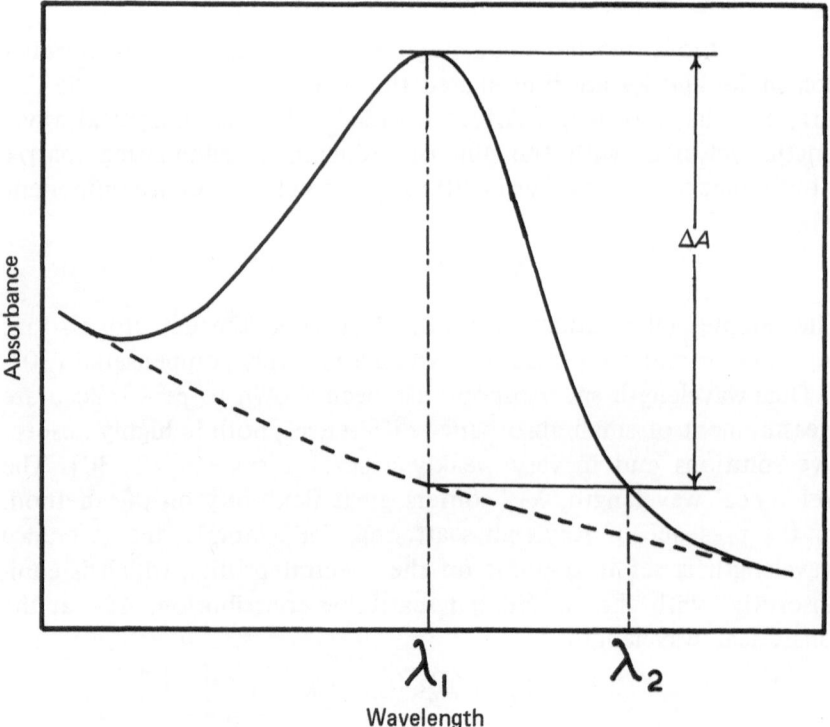

Fig. 11.4 *The measurement of absorbance in dual-wavelength spectrometry. Redrawn from Shibata [2].*

In cases where two defined components X, Y overlap, the spectrum of Y may be deconvoluted or suppressed by selecting, for λ_1 and λ_2, wavelengths which correspond to equi-absorptive points on each side of a peak maximum in the unwanted component. Under these conditions, the difference absorbance for Y must be zero:

$$\Delta A = A_Y^{\lambda_1} - A_Y^{\lambda_2} = 0$$

Thus the difference absorbance for component X will be independent of Y (Fig. 11.5):

$$A_T^{\lambda_1} = A_X^{\lambda_1} + A_Y^{\lambda_1} \qquad \text{at } \lambda_1$$

$$A_T^{\lambda_2} = A_X^{\lambda_2} + A_Y^{\lambda_2} \qquad \text{at } \lambda_2$$

and

$$\Delta A = A_X^{\lambda_1} - A_X^{\lambda_2}$$

$$= \Delta \epsilon_X bc.$$

Dual-wavelength spectroscopy can, in this example, be seen as a special case of difference spectroscopy. Various modifications of this approach to spectral suppression have been proposed, using analogue absorbance multipliers [2]. A method proposed for examining complex formation is to use an isosbestic point as a reference wavelength, in order to increase measurement accuracy and precision [2].

In dual-wavelength spectroscopy the option to use each wavelength independently is available. This feature has been used to monitor two reactants at their respective optimal absorption maxima [2]. However,

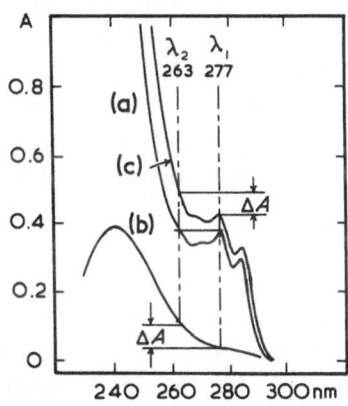

Fig. 11.5 *The absorption spectra of (a) isophthalic acid (10% w/v), (b) terephthalic acid (0.5% w/v), and (c) a mixture containing both compounds in the same concentrations. Redrawn from Porro [30].*

the more common mode is to measure the difference absorbance at two wavelengths, one set to the analyte and the other to a reagent which is consumed during reaction. This method permits the sensitivity to be readily increased [2].

If the two monochromators are set a small spectral interval apart ($\Delta\lambda \approx 1-4$ nm), the first derivative spectrum, $dA/d\lambda$, can be readily obtained by scanning the monochromators in tandem. Moreover, the first derivative of absorbance at fixed wavelength can also be monitored for reaction kinetics. However, the higher derivatives are not accessible by this method, as discussed in Section 11.1.

Although dual-wavelength spectroscopy offers a number of intriguing measurement possibilities in UV–VIS spectroscopy, the cost of instrumentation has precluded its widespread use. The method appears to have found wide application for high sensitivity and high accuracy measurements in complex biological systems, particularly in plant photochemistry.

11.3.1 *Future perspectives*

The advent of rapid scanning spectrometers based on the linear photodiode array or the TV vidicon tube, coupled with powerful microcomputers for data storage, manipulation and presentation, opens up new perspectives in analytical spectrometry. In addition to the capability to generate second- and higher-derivative spectra, difference spectra and log spectra, these devices can be used for matrix least-squares deconvolution of defined, multicomponent mixtures using either zero-order or derivative spectra. The establishment of digital archives of spectra, coupled with rapid archive retrieval routines based on derivative spectra, will open up another phase of development in UV–VIS spectroscopy. Rapid scanning detectors have also shown significant potential for enhanced detection capability in HPLC, where transformation of captured spectra to their second or higher derivative offers a more highly characteristic profile for identification [31].

11.4   **Densitometry**

Accessories for the measurement of the optical density of solid samples such as photographic film, TLC plates or electrophoresis gels are available for some spectrometers. As with all accessories, the spectrometer-plus-attachment is less satisfactory than a custom-built densitometer, particularly because the sample must be mounted vertically, it is difficult to align the beam through the region of interest, and a densitometer will have microscope optics allowing a higher beam intensity and smaller spot size to be used. On the other hand, the spectrometer has some advantages over densitometers that only operate with a measuring beam of white light, for the optimum wavelength for a particular measurement can be selected in order to minimize the effects of light-scattering, and UV-absorbing compounds can be measured directly without the necessity of staining them. The spectrometer can also measure the complete absorption spectrum of a particular spot, though the high background absorption and scatter will probably mean that the spectrum is of poor quality.

In general, the sample is mounted vertically in the measuring beam of the instrument, and the accessory will have some means of moving the sample horizontally so that it is scanned along its length. This motion requires a spectrometer with a large sample compartment and widely separated measuring beams, and consequently not all instruments can accommodate such an accessory.

Two common applications of densitometry in chemical and biochemical laboratories are the measurement of spots on a TLC plate, which may be coloured or stained, and the measurement of bands in electrophoresis gels. The latter may be in the form of a 'slab' – a sheet 1–2 mm thick – or a 'tube', which is a gel cylinder about 6 mm in diameter. These are usually stained with a blue dye, although in some cases direct measurements of the absorption of the bands can be made. The task is to measure the optical density of the separated spots or bands and hence the concentration of the components of a mixture, and to identify these components by the location of the spot on the sample. Thus the densitometer scan will consist of a plot of optical density or transmission of the sample against distance from some reference point. These samples are 'difficult' in spectrometric terms since the gels or TLC plates are generally opaque and scatter light badly. It is therefore an advantage to use light of as long a wavelength as possible and the blue dyes used in gel electrophoresis with $\lambda_{max} > 600$ nm are ideal. In some cases, a better approach is to photograph the gel by either transmitted or reflected light and scan the resulting negative. The operation is simple if a Polaroid camera with negative film is used. The loss of resolution caused by this extra process can be compensated by the increased contrast of the negative, and the negative is far easier to mount in the spectrometer. This technique is particularly useful if fluorescent bands are to be measured, the gel being photographed with an appropriate exciting wavelength while the camera is fitted with a filter to block the exciting light.

11.4.1 *A typical densitometer accesssory*

Fig. 11.6 shows a typical densitometer accessory consisting of a carriage mounted on rails and driven by a stepping motor. This carriage can carry one of a variety of sample holders, which can accommodate photographic film or plates, strips cut from slab gels contained in a fused silica trough or tube gels in a stoppered silica tube. Provision is made for a movable mask to define the height of the measuring beam as it passes through the specimen and there are two stationary vertical slits which define the width of the measuring beam and hence the spatial resolution of the scanned spots. A clear distinction must be made between the monochromator slits, which define the *spectral*

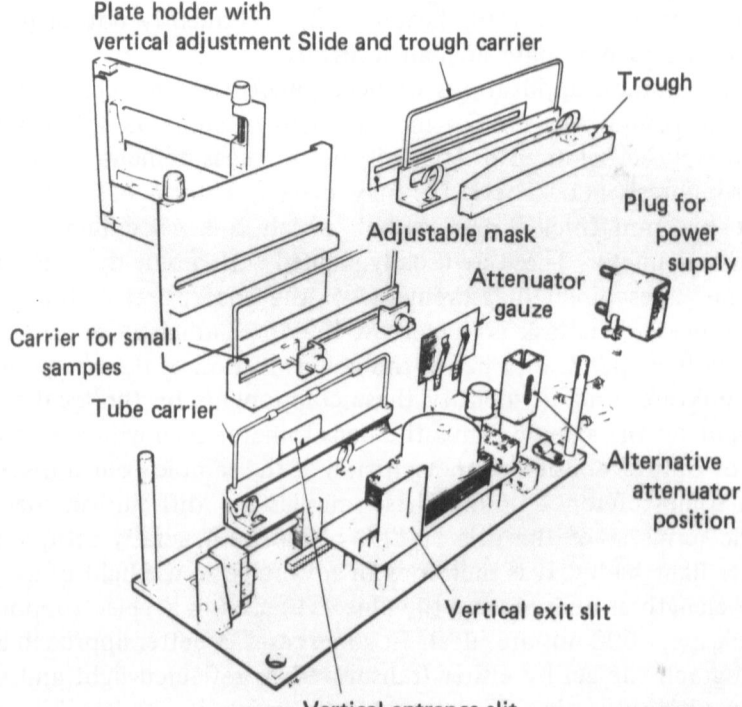

Fig. 11.6 *A densitometer accessory supplied for use with the Pye Unicam SP 8 spectrometers, showing the basic unit with tube carrier in place, and with the alternative carriers for large and small plates and for slab gel slices. The measuring beam passes through the two vertical slits and the sample while the reference beam passes through the attenuator at the right-hand side of the unit. Reproduced by permission of Pye Unicam Ltd.*

resolution of the measurement, and the accessory slits which determine the ability of the instrument to distinguish between closely-grouped spots or bands on the sample. Fig. 11.7 illustrates the test-piece supplied with the accessory and the trace recorded from it with a 0.1 mm slit. The horizontal lines on the trace represent 0 and 1 T, while the relationship between the movement of the sample and the movement of the chart is such that the horizontal scale is magnified about twice. The pen has reached the T = 0 line when scanning over the broad opaque regions at the centre of the test piece, but with lines less than 0.5 mm wide, the relative rates of scanning speed and pen response have resulted in peak heights that are less than their true values. Similarly, at the clear zones between opaque blocks the pen fails to reach the true transmittance value of unity. Both problems are illustrated when the fine doublet of lines is scanned, for the lines neither show their correct transmittances nor are they properly resolved. Thus just as when recording absorption spectra, in order to obtain a high-resolution trace, a small slit-width must be used with a

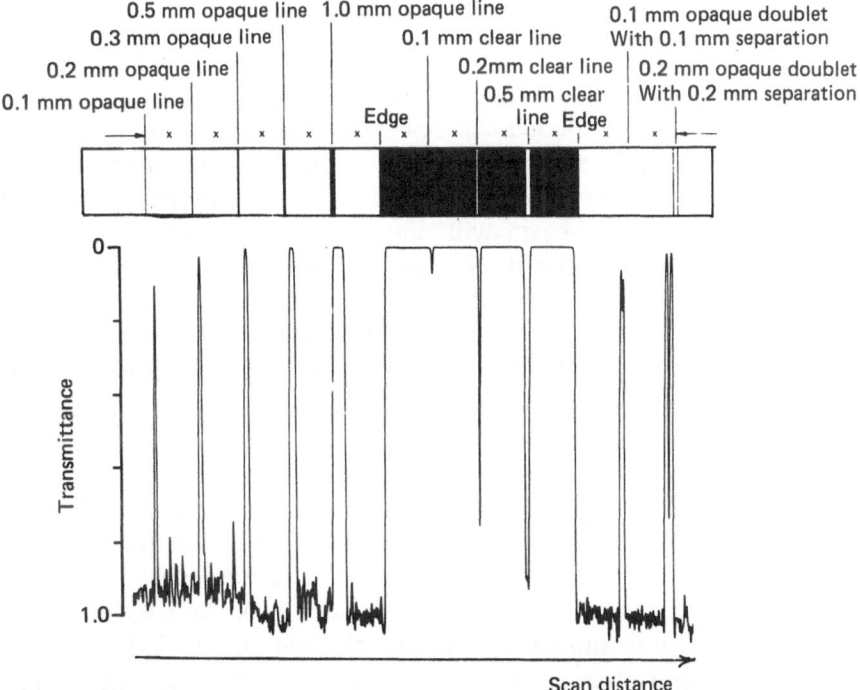

Fig. 11.7 *Diagram of the test piece supplied with the Pye Unicam densitometer and a trace obtained from it using a vertical slitwidth of 0.1 mm.*

scanning speed that is sufficiently slow to allow the recorder pen to keep up with the scan and show a true transmittance value.

Attenuation of the beam by the accessory slits and the sample itself means that the reference beam must also be strongly attenuated if an approximate beam balance is to be obtained. In the illustrated accessory this can be done by means of blackened gauze held in a spring clip or by the use of a 10 mm cell or block of suitable absorbing material.

11.4.2 *Sample handling*

Rigid samples such as photographic film and TLC plates present few problems. The region of interest should be identified and the start and finish points of the scan marked with black ink or by scoring the sample. If there are several tracks to be scanned on the same sample, these marks should be made across the sample exactly perpendicular to the scan direction. Softer specimens like polyacrylamide gels are more difficult to handle because of their elasticity and because they will shrink if allowed to dehydrate. Some flexible specimens can be supported simply by sandwiching them between glass plates, but the softer gels must be immersed in a buffer solution of composition similar to that used in running them. First mark the start and finish points; it is prudent at this stage to measure the distance between these points so that the gel can be chceked for distortion before it is measured. Slab gels must be cut very accurately into strips using a ruler and razor blade, for if the edge is not exactly parallel to the band on the gel, the beam will run off the edge of the band as it scans. The strip is put into the trough, which is then filled with buffer to eliminate bubbles and air pockets. Tube gels must be pushed into the sample tube with the minimum distortion of the specimen. The tube is then filled with buffer, with tilting and tapping to displace the bubbles, and then carefully stoppered.

11.4.3 *Using the accessory*

Mount the accessory in the instrument and check the direction of sample traverse by the motor. Set the carrier to its starting position, set the monochromator to 550 nm, open the monochromator slits wide and with a small piece of white card, check the passage of the measuring beam through the accessory. Put the sample in the appropriate holder and place the holder on the carriage, select a mask that will give the maximum height of beam compatible with the width of the band on the sample, and put the mask in position, taking care to ensure that it is horizontal. Select the vertical accessory slits so that these are also as wide as possible but able to resolve the smallest feature on the sample. Check with the white card that light is passing through both slits and the sample.

Set the monochromator slitwidth to the operating value; in general, this can be quite wide since high measuring-beam intensity is required rather than high spectral resolution. Move the carriage so that the beam passes through a clear part of the sample and adjust the reference beam attenuator to bring the pen to unit transmittance. If gauze screens are used, this may require the use of two or more screens to obtain the necessary reduction in intensity. Adjustment of these is then a very delicate matter for very small movements of one screen relative to another can make great changes as the holes go in and out of registration. Next move the carriage to the densest part of the sample and select the appropriate transmittance or absorbance scale. Having chosen a reasonable sample scanning speed and returned the carriage to the starting position, the sample is ready to be measured.

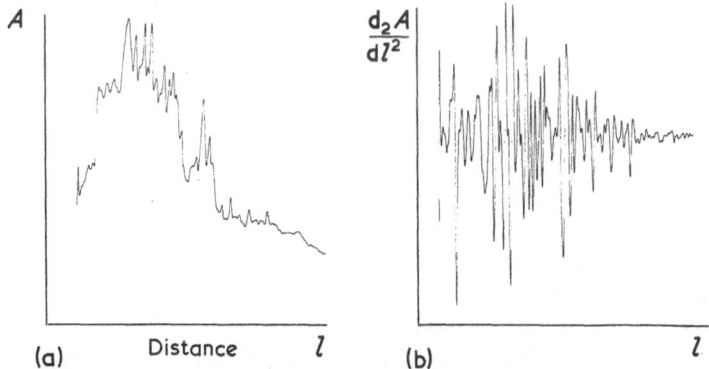

Fig. 11.8 *Enhancement of a densitometer trace by plotting the second derivative: (a) conventional scan of an autoradiograph film, showing much fine detail superimposed on a variable background; (b) second derivative of (a) showing the elimination of the background and sharpening of the peaks. Reproduced by permission of Pye Unicam Ltd.*

If the instrument can generate derivative spectra, the densitometer trace may be improved by plotting it as the second derivative (Fig. 11.8). This will eliminate slow changes in the background and sharpen small peaks and shoulders.

References

1 Chance, B. (1954), *Rev. Sci. Instrum.*, **22**, 634.
2 Shibata, S. (1976), *Angew. Chem. Int. Ed. Engl.*, **15**, 673.
3 Giese, A.T. and French, C.S. (1955), *Appl. Spectrosc.*, **9**, 78.
4 Singleton, F. and Collier, G.L. (1956), *Brit. Pat.*, **760**, 729.
5 Collier, G.L. and Singleton, F. (1956), *J. Appl. Chem.*, **6**, 495.
6 Olson, E.C. and Alway, C.D. (1960), *Anal. Chem.*, **32**, 370.
7 Fell, A.F. (1980), *UV Spectrom. Grp Bull.*, **8**, Part 1, 5.
8 Morrey, J.R. (1968), *Anal. Chem.*, **40**, 905.
9 Butler, W.L. and Hopkins, D.W. (1970), *Photochem. Photobiol.*, **12**, 439, 451.
10 Talsky, G., Mayring, L. and Kreuzer, H. (1978), *Angew. Chem. Int. Ed.*, **17**, 785.
11 Ishii, H. and Satoh, K. (1982), *Fres. Z. Anal. Chem.*, **312**, 114.
12 O'Haver, T.C. and Green, G.L. (1976), *Anal. Chem.*, **48**, 312.
13 Hawthorne, A.R. (1980), *Am. Ind. Hyg. Assoc. J.*, **41**, 915.
14 Savitzky, A. and Golay, M.J.E. (1964), *Anal. Chem.*, **36**, 1627.
15 Steiner, J., Termonia, Y. and Deltour, J. (1972), *Anal. Chem.*, **44**, 1906.
16 Traveset, J., Such, V., Gonzalo, R. and Gelpí, E. (1981), *J. Chromatogr.*, **204**, 51.
17 Traveset, J., Such, V., Gonzalo, R. and Gelpí, E. (1980), *J. Pharm. Sci.*, **69**, 629.
18 Fell, A.F. (1978), *Proc. Anal. Div. Chem. Soc.*, **15**, 260.
19 Fell, A.F. (1982), *Proc. Symp. Analysis of Steroids*, Eger, Hungary (Ed. S. Görög), Elsevier, Amsterdam, p. 459.
20 Davidson, A.G. and Elsheikh, H. (1982), *Analyst*, **107**, 879.
21 O'Haver, T.C. (1979), *Clin. Chem.*, **25**, 1548.
22 Gill, R., Bal, T.S. and Moffat, A.C. (1982), *J. Forensic Sci. Soc.*, **22**, 165.
23 Fell, A.F. (1983), *Trends Anal. Chem.*, **2**, 63.
24 Görög, S. (1968), *J. Pharm. Sci.*, **57**, 1737.
25 Chafetz, L., Tsilifonis, D.C. and Riedl, J.M. (1972), *J. Pharm. Sci.*, **61**, 148.
26 Ruckpaul, K., Rein, H., Ballou, D.P. and Coon, M.J. (1980), *Biochim. Biophys. Acta.*, **626**, 41.
27 Ichikawa, T. and Terada, H. (1979), *Biochim. Biophys. Acta*, **580**, 120.
28 Willard, H.H., Merritt, L.L. and Dean, J.A. (1974), *Instrumental Methods of Analysis*, Van Nostrand, New York, p. 94.
29 Donbrow, M. (1967), *Instrumental Methods in Analytical Chemistry*, Vol. II, *Optical Methods*, Pitman, London, p. 130.
30 Porro, T. (1972), *Anal. Chem.*, **44**, 93A.
31 Fell, A.F., Scott, H.P., Gill, R. and Moffat, A.C. (1972), *Chromatographia*, **16**, 69.

12 Automated sample handling

12.1 Introduction

Laboratories often meet large, unexpected increases in the demand for a particular analytical determination. In many of these instances it has only been possible to respond to these demands within the required time by recourse to automated systems for sample handling. A close examination of existing practices may reveal advantages to be gained by the automation of sample-handling thus releasing trained staff for more demanding tasks. Some laboratories have the resources to develop and construct their own dedicated systems but most laboratories depend on the purchase of suitable equipment from the fairly wide range of modules which are available. Adaptations to commercial equipment can be made to render it more suited to specific requirements.

Automated sample handling systems can be divided into the following broad categories:

(a) Discrete systems
(b) Centrifugal systems
(c) Continuous-flow systems with segmentation
(d) Continuous-flow systems without segmentation.

This chapter contains brief descriptions of the first two types and a more detailed account of continuous flow systems.

Discrete systems are those which closely reproduce the classical operations of pipetting, reagent addition, treatment and measurement by electro-mechanical methods.

Centrifugal systems are those in which samples and reagents are brought together and transferred to an optical cell on a rapidly rotating turntable. Absorbance measurements are made whilst the cells are rotating and resolution of the individual signals is carried out electronically.

Continuous-flow (CF) systems are those through which a continuous flow of liquid is pumped and into which the sample solutions are successively inserted. Following subsequent addition of reagents and diluents and a variety of treatments, the stream passes through the flow cell of an appropriate detector.

CF automated analysis was introduced by Skeggs in 1957 [1] and commercially exploited at first by Technicon as the Auto Analyzer (R). Since its introduction, the Auto Analyzer has undergone a number of improvements leading to standard laboratory systems which can achieve sampling-rates of up to 60 per hour and advanced multi-analysis clinical systems which can perform up to 20 simultaneous determinations at rates of up to 150 samples per hour. These developments are all based on air-segmented systems and are due to the increased understanding of the scientific principles involved [2–10].

During the later period of development of air-segmented systems, there has been a fairly rapid growth in the development of non-segmented CF systems, under the general description of *flow injection analysis* (FIA). In this technique, samples are introduced into the flowing stream by means of the injection loop of a liquid sampling valve. Thus a fixed volume of the sample passes through the system as a discrete slug as opposed to being distributed over a series of liquid segments corresponding to a given time of sampling, as in the segmented systems. This type of system will be described in more detail later.

12.2   Air-segmented continuous-flow systems

12.2.1   *Introduction*

Air-segmented CF analysers are the most widely used and versatile systems. Hundreds of papers have been published on the theory and applications of this technique, but there may still be a number of laboratories where the advantages have not been fully appreciated. The main aim of any automated system must be to maximize output. Many determinations require strict adherence to a rigorous timing schedule, and while humans may be fallible in this respect, an auto-analyser excels. Some determinations entail the handling of reagents which, even if not hazardous, may constitute a nuisance in the laboratory and the use of a closed system with appropriate waste-disposal can often be beneficial. CF systems certainly act in the interests of health and safety, for they are capable of handling a range of hazardous liquids. However, any leakage or blockage of the system may lead to a burst, which could be dangerous unless appropriate safety measures are taken.

A schematic layout of a typical laboratory system is shown in Fig. 12.1 and with more detail in Fig. 12.2.

The sampler has to be capable of taking up solutions from successive sample cups for a pre-determined period of time, followed by an intermediate sampling of appropriate wash liquor, also for a pre-determined length of time. These times must be accurately reproducible and the time of passage of the probe between sample and wash must be minimized so as to limit the amount of air drawn into the system during this movement. A small air bubble does, however, help to sweep the sample line. Most samplers have a rotary sample plate, with peripheral holes to accommodate the plastic sample cups. Some have two concentric rings of holes and twin probes as a means of carrying out two simultaneous determinations. It is worth obtaining the type of plate with holes to take either 2 or 5 ml cups since certain applications call for high sampling rates when the larger cups become essential.

Some samplers are timed by means of rotating cams and microswitches, while others use electronic timing. Cams provide a very reliable and robust method but lack the total flexibility of electronic systems. It is possible to use external electronic timers to replace the use of cams. A correct choice of sampling and wash times is vital for optimal operation of the system, details of the method of deriving minimum sampling times will be found in the references. Wash times should normally be just sufficient to give adequate peak separation. The chosen times can be validated by running three 'low' samples, followed by three 'high' samples and a further three 'low' samples. Any unacceptable carry-over will be evidenced by the first 'high' peak being pulled down by the preceding 'low' and vice versa.

Some samples may produce a sediment whilst awaiting their turn to be sampled. Agitator attachments are available which will enable the sample to be homogenized during sampling. The sampling of viscous liquids can give rise to carry-over problems in the non-segmented sample line. One sampler exists which can be made to operate in a 'pecking' mode so that air-segmentation is introduced at the sample probe. Conventional samplers are totally unsuitable for

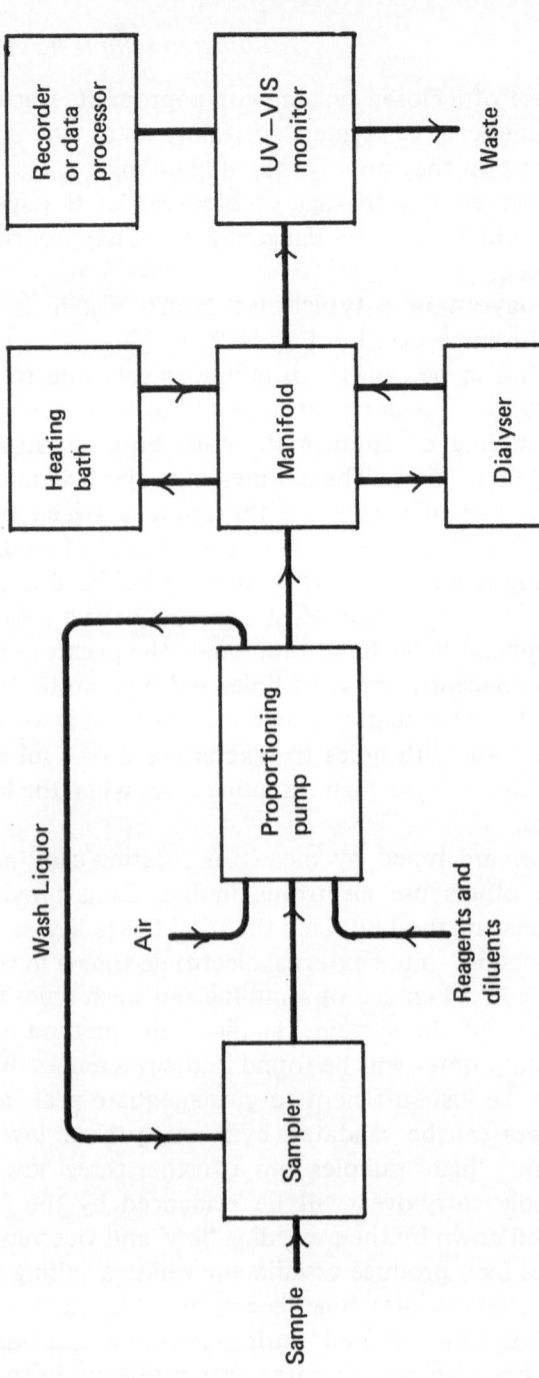

Fig. 12.1 *Schematic diagram of an air-segmented continuous-flow analysis system.*

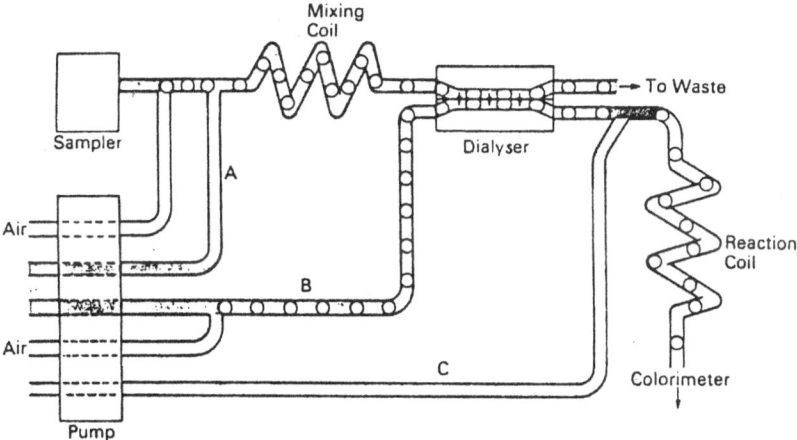

Fig. 12.2 *Schematic diagram of an air-segmented continuous-flow analysis system using sample separation by dialysis. From Snyder et al. [10].*

unstable samples which may have to reside on the turntable for an hour or so awaiting their turn. Newer types of autosampler are available which prepare the sample solution just prior to the actual sampling. A sampler is available for the automatic sample preparation from solids such as tablets and capsules.

12.2.2 *Pump systems*

The heart of any system is the multi-channel peristaltic pump. Its function is to meter the sample and reagents or diluents and to generate the essential regular, correct bubble pattern. The ideal bubble is slightly longer than its diameter and the optimum bubble frequency is determined by the type of manifold [5]. The Technicon Pump III uses an 'air-bar' which is designed to inject the air at an optimum rate for a system based on 2.0 mm internal diameter (i.d.) components. Other systems rely on a correct choice of relative liquid and air pumping rates to achieve the best bubble patterns.

Pumps are available which will accommodate as many as 28 pump tubes in a single layer. Earlier designs of pumps have their tubes mounted in two layers; the actual capacity of the pump, rather than the rated capacity, is often dependent on the particular combination of pump tubes required. The pumping rate is dependent upon the actual diameter of the pump tube. The sample, reagent and diluent *volumes* used in the related manual method must be converted into corresponding *flow rates.* For this purpose colour-coded pump tubes are available covering a wide range of pumping rates. They are available in Tygon (for normal liquids), Solvaflex (for certain solvents) and Acidflex (for strong acids). Silicone-rubber pump tubes are a more recent addition; they are relatively expensive but can be used with liquids which are not suited to the other materials. 'Flow-rated' pump tubes have a closer tolerance on the pumping rate than the standard range but are not available in silicone-rubber. The extra cost of flow-rated tubes is often worthwhile for more exacting analyses. Pumping difficulties may be experienced if attempts are made to utilize the smallest of the available range of pump tubes. The exact 'cut-off' size will depend on the choice of pump.

There is one problem in the unattended shutdown of a pump, which is the inability to automatically release the roller pressure. If the pump is stopped for any length of time without relieving the pressure, the tubes become permanently damaged. There are two ways to avoid the wastage of expensive reagents without endangering the pump tubes on shutdown: one is the use of special valves which may be operated manually or automatically to switch the pumping from reagents to wash liquid. The other method is to switch the pump into a stand-by mode in which it continues to operate intermittently. This latter facility is useful when a system may be needed again at short notice.

12.2.3 *The manifold*

This is the assembly of glass fittings, mixing coils and interconnecting tubing which forms the arterial liquid-handling system. Two ranges of components for standard laboratory systems are available, termed System I and System II. Many laboratories have retained their early System I equipment and replacement modules and fittings can still be obtained. The i.d. of these components is about 2.3 mm. The System II components are generally smaller, with an i.d. of 2.0 mm. Lower pumping rates are used with a saving in reagent consumption, while better wash characteristics are obtained thus allowing an increased throughput of samples.

System I manifolds are normally constructed on a tray mounted on the pump. System II manifolds can be constructed in, and on, a special manifold module or 'cartridge'. Usually it is not possible to devote an autoanalyser to a single determination and regular changes of function are necessary. Although, with practice, System I manifolds can be changed over fairly conveniently, the storage of spare manifolds can be a problem. The System II cartridges are much more convenient, in this respect, and they can also include the smaller System II heating baths and dialysis units. It is advisable to keep the manifold and pump tubes assembled, so spare end-blocks for the pump are useful; the tubes can be taped to the end-blocks to prevent them getting tangled during storage.

The standard transmission tubing is 1.6 mm i.d. Tygon. Tubing in Solvaflex, Acidflex and silicone-rubber is also available. Polythene and glass are thought to offer better wash characteristics but are obviously more difficult to work with from the point of view of bending and joining. Various 'bends' are available in glass transmission tubing, which ease the problem of constructing all-glass manifolds. Joints in the flexible tubing are made using the range of small connectors (nipples) which are available in plastic, glass and stainless steel–platinum. With Tygon and Solvaflex, sleeved butt-joints can be made using cyclohexanone to weld tubing and sleeving together.

12.2.4 *Heating*

Owing to the precise reproducibility of these systems it is generally held to be unnecessary to heat the reaction mixture for as long as it takes to achieve 100% completion of the reaction. This is generally true, but a word of caution is necessary. In some instances, unless the reference solution is identical in composition to the samples, the rates of reaction can differ and erroneous results may be obtained. Such problems can be minimized by allowing the reaction to go to completion.

System I heating baths are large separate modules, accommodating coils of up to 25 m in length. System II heating baths are small, sealed units which can be accommodated in the manifold cartridge unit. However, System II heating baths are often too small to give long delay times and the larger type of bath may have to be utilized. The size of coil required for a given residence time can be calculated from the total flow rate and its i.d. It is usual to categorize the smaller System II coils by volume and the larger coils by length.

12.2.5 *Dialysis*

Dialysis has been used as a convenient means of separating the test substance from irrelevant high molecular weight material. As with heating baths, the larger System I modules are still available. System II units are much smaller and can form a part of the manifold. A variety of membranes is available to suit particular applications.

Fig. 12.2 shows a typical system. The proportion of the analyte of interest transferred across the membrane is quite low since the dialysis unit cannot be used in a counter-current mode and since some of the dialysers are quite small. However, this can be an advantage as it may avoid the need for a dilution stage. The flow rate of donor and recipient streams should be matched as closely as possible which may call for the use of a slightly higher pumping rate for the recipient stream due to back pressure from the rest of the system.

12.2.6 *Detectors*

CF analysis was initially designed for colorimetric detection, but any method of detection can be used which will give a measurable signal and for which a suitable flow-cell can be constructed. Various colorimeters are available mostly based on interference filters. One colorimeter offers at least four channels, operating from a single light-source. For complete versatility, the use of a double-beam UV–VIS detector is preferable to a colorimeter. Any stable spectrometer may be used and a range of flow cells is available for most instruments.

It is usual to 'debubble' the stream, before drawing a portion through the cell although debubbling has an adverse effect on sampling rates. The technique of 'bubble-gating' is now introduced whereby the air is permitted to pass through the cell and the effects filtered out electronically [11, 12].

12.2.7 *Data processing*

Owing to the prolific output of data from CF systems, the time devoted to manual calculations can compare very unfavourably with the analysis time, and electronic data processing becomes imperative. Microprocessor systems are now available which are capable of controlling the system and monitoring its performance in addition to the generation of results. Under the heading of data processing one could include the technique of 'curve regeneration'. It is normal in CF analysis to sample for a sufficient time to allow the response to effectively reach the maximum value, the so-called 'steady state' condition. This can limit the throughput of a system. The curve regenerator [13, 14] allows the use of reduced sampling times by continuously calculating the height of the steady-state value from the rate of increase in the response. Increases of up to 50% in throughput can be obtained though for precise work, such systems must be carefully validated.

12.3 Flow injection analysis

Until relatively recently it was assumed that segmentation was an essential feature of CF analysis, but FIA is a viable alternative. In this technique the sample is inserted as a slug into a flowing stream of suitable carrier liquid containing the reagent by means of a rotary valve with an injection loop. A schematic system is shown in Fig. 12.3. FIA was developed independently by Stewart *et al*. in the USA and by Ruzicka and Hansen in Denmark. The former chose tubing with an i.d. of less than 0.5 mm. Since pumping pressures are in the region of 100–500 psi an HPLC type of pump is required. The latter group use 0.5–1.0 mm tubing and a peristaltic pump is suitable. Commercial systems are available, based on both principles. Betteridge [15] has written a comprehensive report on this technique.

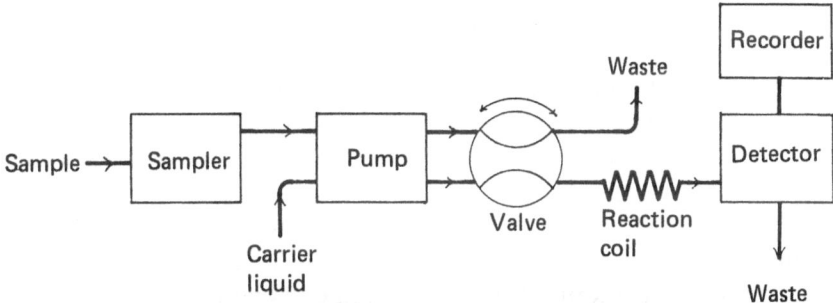

Fig. 12.3 *Schematic diagram of a flow injection analysis system.*

The advantages of FIA are speed (up to 300 samples per hour), simplicity, and reduced cost of the apparatus. The very simplicity of this technique constitutes its main disadvantage. Very high sampling rates can be achieved for the simpler analytical procedures but the technique cannot be applied to the more lengthy and elaborate operations, for which the air-segmented approach is better suited. There is still a lot of scope for innovation in both techniques.

12.4 Other CF techniques

Solvent extraction is a well-established technique in CF systems, using ordinary mixing coils followed by the appropriate separator fitting. Fittings, with special plastic inserts, are available according to whether the organic phase is light or heavy and which of the two separated phases is required. The extraction solvent is not always suitable for direct injection on to an HPLC column. In these cases an EDM (R) (evaporation to dryness module) is available in which the sample is continuously evaporated on to a moving 'wire' and then re-dissolved into an alternative solvent.

The high cost of specific enzymes has often precluded their use as continuously pumped reagents, but immobilized enzymes, coated on the inner surface of CF fittings, can be used for a variety of determinations including glucose (glucose oxidase) and penicillin (β-lactamase). Both air-segmented and FIA systems have been used.

A stream splitter can be used to divide a sample stream equally between two manifolds. By this means, two separate determinations can be carried out using a single sample probe. If the two manifolds are identical, an essential reagent can be substituted in one manifold to make this stream an effective 'blank'. By leading the two streams to the respective cells in a twin-channel detector, simultaneous correction can be obtained by subtracting the two signals. It is necessary to accurately synchronize the two channels: this is easily accomplished by sampling a liquid which will give an equal response in both channels. By adjusting the lengths of transmission tubing feeding the two flow cells, a null signal will be obtained when the two channels are in phase. The mercury-catalysed 'imidazole' assay for β-lactam antibiotics can be conducted in this way by simply omitting mercuric chloride from the 'blank' channel. A very wide range of formulated products can be assayed without interference from excipients.

A variety of switching-valve arrangements has been used to monitor the progress of tablet dissolution-rate tests by linking up to six dissolution vessels to a CF analysis system.

In this short chapter it has only been possible to give an outline sketch of CF analysis, and whole books [16–19] have been written on the subject. It is hoped that a few appetites may have been whetted and that those previously unfamiliar with this technique be persuaded to dig deeper.

References

1 Skeggs, L.T. (1957), *Amer. J. Clin. Pathol.*, **28**, 311.
2 Aris, R. (1956), *Proc. Roy. Soc. A*, **235**, 67.
3 Sutter, A., Gardanier, S.A. and Spooner, G.H. (1970), *Amer. J. Clin. Pathol.*, **54**, 341.
4 Snyder, L.R. and Adler, H.J. (1976), *Anal. Chem.*, **48**, 1017.
5 Snyder, L.R. (1976), Advances in Automated Analysis *7th Technicon Inter-*

national Congress. Vol. 1, Technicon Instrument Corp., Tarrytown, New York, p. 76.

6 Thiers, R.E., Cale, R.R. and Kirsch, W.J. (1967), *Clin. Chem.*, **13**, 145.

7 Pennock, C.A., Moore, G.R., Collier, F.M. and Barnes, I.C. (1973), *Med. Lab. Tech.*, **30**, 145.

8 Thiers, R.E., Reed, A.H. and Delander, K. (1971), *Clin. Chem.*, **17**, 42.

9 Walker, W.H.C. and Andrew, K.R. (1974), *Clin. Chim. Acta.*, **57**, 181.

10 Snyder, L.R., Levine, J., Stoy, R. and Conetta, A. (1976), *Anal Chem.*, **48**, 942A.

11 Neeley, W.E., Wardlaw, S.C. and Swinnen, M.E.T. (1974), *Clin. Chem.*, **20**, 78.

12 Neeley, W.E., Wardlaw, S.C., Yates, T., Hollingsworth, W.G. and Swinnen, M.E.T. (1976), *Clin. Chem.*, **22**, 227.

13 Walker, W.H.C., Townsend, J. and Kean, P.M. (1972), *Clin. Chem. Acta.*, **36**, 119.

14 Caryle, J.E., McLelland, A.S. and Fleck, A. (1973), *Clin. Chem. Acta.*, **46**, 235.

15 Betteridge, D. (1978), *Anal. Chem.*, **50**, 832A.

16 Furman, W.B. (1976), *Continuous Flow Analysis*, Marcel Dekker, New York.

17 Coakley, W.A. (1981), *Handbook on Automated Analysis*, Marcel Dekker, New York.

18 Foreman, J.K. and Stockwell, P.B. (1975), *Automated Chemical Analysis*, Ellis Horwood, Chichester.

19 Ruzicka, J. and Hanson, E.H. (1981), *Flow Injection Analysis*, Wiley, New York.

13 The maintenance of instruments

13.1 Introduction

A UV–VIS spectrometer is capable of precise measurements only when it is carefully maintained. This chapter outlines a check and maintenance programme that is within the capability of all users. It is not possible to give detailed instructions for every spectrometer in use today, but consultation of the manufacturer's handbook should fill in the specific details of the procedures.

The fundamental maintenance routine should involve: (a) elementary maintenance aimed at preventing erroneous results; (b) a regular instrument check routine; (c) diagnosis of both progressive and sudden faults by monitoring the results from successive check routines; and (d) the correction of faults, where this is possible.

13.2 Environmental and safety considerations

13.2.1 Environment

Siting modern spectrometers is less demanding than it was for their older counterparts, which required specially constructed tables to ensure the stability of the optical bench. Even so, all spectrometers should be sited in the best environment that is available. The versatility of absorption spectroscopy means that some instruments will be used in highly unsuitable environments, and these instruments must be guaranteed a maintenance programme that is more extensive and frequent than the norm.

Ideally the spectrometer should be sited in an area which is dust-free and away from direct sunlight and draughts. The original concept of a special instrument room was later thought to be unnecessary; however, it is now coming back into favour, and is well worth adopting if space is available. It should be noted that providing bench space for the instrument alone is not sufficient — if there is no room available for flasks, cells, etc., they will undoubtedly be placed on top of the instrument with inevitable spillages. Easy access to the back and sides of the instrument makes maintenance easier, and if this is not possible, ensure that the bench is sufficiently wide for the instrument to be turned round.

Services should amount to more than a single mains socket, for accessories – even if only a soldering iron – will require mains electricity. Purging the instrument to remove oxygen is necessary before meaningful results can be obtained below 200 nm. This is best achieved by using nitrogen gas, either boiled off from liquid nitrogen or using cylinder nitrogen which is adequately filtered. A time switch can be useful in saving operator time and avoiding unnecessary running of the source lamps, which have limited lifetimes.

13.2.2 *Safety*

All of the hazards associated with electrical equipment are likely to be encountered. Only persons designated as 'competent' should be allowed to investigate faults. The instrument cover should only be removed by such a person, and it should remain fixed in place at all other times. While modern electronics operate at a few volts, the dynode supply for the photomultiplier tube may be over 1000 V.

Beyond the electrical hazards, two other features of UV–VIS spectrometers are potentially dangerous:

(a) The generation of ozone by the UV source lamp. This toxic gas is usually only encountered in older instruments, but it may be present whenever a UV lamp is running and not all people can detect it by smell. It must be removed from the lamp house and vented outside the work area.

(b) UV radiation damages the eyes and causes irritation of the skin. Operation of the UV lamp without its cover in place should only be allowed during maintenance periods, and then adequate eye protection must be worn by all personnel if they are liable to be exposed to either direct or reflected radiation. The absorption characteristics of a variety of lens materials giving differing degrees of UV protection are given in Fig. 13.1, but goggles opaque to wavelengths below 350 nm are advisable. Hughes [1] gives a full account of UV hazards and the precautions to be taken.

13.3   Routine checks

13.3.1 *General principles*

Routine spectrometer checks should be carried out by the person most familiar with the instrument – namely the user. The extent and

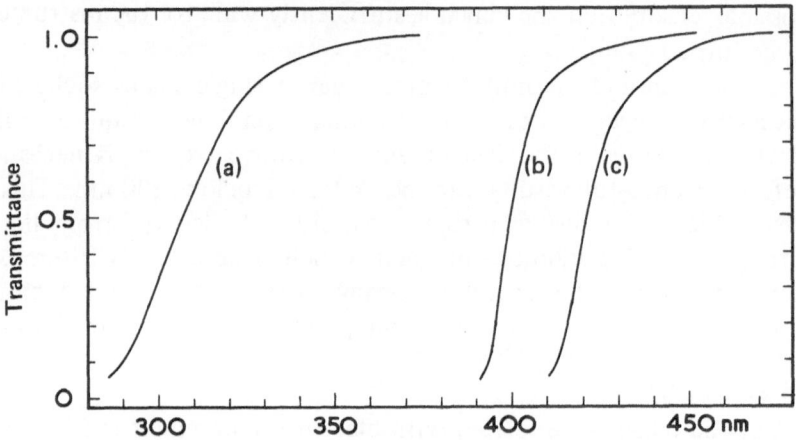

Fig. 13.1 *Transmission spectra of protective goggle lenses: (a) Toughened glass, (b) polycarbonate plastic, (c) 'Blak—Ray' UV safety lenses.*

frequency of the checks must be determined by one's experience of the instrument and the degree and nature of its use. Having decided upon a check programme, it must be strictly adhered to, irrespective of whether every day, every week or every month is considered to be an appropriate interval between checks. It is advisable to design a proforma which can be readily understood and carried out by the operator performing the checks. An example is shown in Fig. 13.2 which includes checks on the wavelength and absorbance accuracy of the instrument and its stray-light level. These tests will be described in detail in the following sections. The instrument operating parameters such as slitwidth, scan speed and direction, light source, etc., should be specified for each test and each instrument.

The completed proforma should be countersigned by a responsible person who will consider the results, decide upon a course of action and ensure that it is carried out. The check report sheets should be filed for future reference and, in addition, it is advisable to plot the data on a control chart. Edisbury [2] discusses the statistical analysis of such a chart, but even if the chart is assessed in the most simplistic way, it will provide an instantaneous view of how the instrument is 'ageing'.

The tests described in the following sections should be regarded as an absolute minimum and, if possible, should be expanded by the inclusion of other checks such as baseline flatness, noise, resolution, lamp output, etc., that are applicable to the instrument and the type of measurement that it is used for.

UV–VIS Spectrophotometer Check

Instrument _____ Date _____

Weight of potassium dichromate_____ g

Nominal wavelength (nm)	Observed wavelength (nm)	Absorbance	Blank absorbance	Corrected absorbance	$A/E_{1cm}^{1\%}$ nominal	$A/E_{1cm}^{1\%}$ observed
350					107.2	
313					48.8	
257					145.4	
235					125.0	

Holmium oxide filter wavelength accuracy check

Nominal wavelength	Observed wavelength
279.4	_____
360.9	_____
453.2	_____

Stray-light determination

Stray light at   nm using   =

Checked by_____ Date_____

Comments

Fig. 13.2 *An example of a check sheet for routine spectrometer testing.*

13.3.2 *Wavelength calibration*

When the most precise wavelength calibration is required, the emission lines of a gas discharge lamp should be measured. A range of suitable lamps is given in Volume 1 of this series ([3], pp. 111–20, 131–2). The lamp should be inserted into the lamp house so that its radiation falls on the entrance slit of the monochromator. Most of these lamps emit UV radiation and so the precautions described above must be observed. Double-beam instruments must be put into single-beam or 'energy' mode so that the output of the lamp can be measured. The instrument is slowly scanned over the wavelength range of interest using minimum slitwidths.

If your spectrometer has a deuterium lamp as UV source, the two prominent emission lines at 486.0 and 656.1 nm can be used as a quick check of the wavelength calibration in that part of the visible region. It is unwise to assume that if these two wavelengths are correct that the UV and longwave visible region are also correct.

For the majority of laboratories, holmium solutions or holmium and didymium glass filters provide a convenient means of checking wavelength calibration which is sufficiently accurate for most purposes. Table 13.1 gives peak positions for typical holmium and didymium filters and Appendix A4 gives the spectra of filters and solutions. While it is not necessary to use all of these maxima, it should be noted that they differ in intensity and band shape, and if only a few maxima are used, they must be chosen with care in order to ensure that the test retains its validity. Holmium glass is preferable to didymium though it is in short supply at the present time. Some instrument manufacturers can supply a piece 12 mm wide which will fit into a cell holder. Larger pieces must be positioned against the cell holder. The instrument is put into absorbance mode with minimum width slits, and slowly scanned past the peaks of interest. Further details of wavelength tests are given in Volume 1 ([3], pp. 111–21, 131–2).

Table 13.1: *Typical wavelengths and tolerances for selected maxima of holmium and didymium glass filters. Note that variations are found between different batches of glass. Taken from Edisbury [2]. See also Appendix A3.*

Holmium glass		Didymium glass	
λ (nm)	$\bar{\nu}$ (cm^{-1})	λ (nm)	$\bar{\nu}$ (cm^{-1})
241.5 ± 0.2	41 410 ± 30	573.0 ± 3.0	17 450 ± 85
279.4 ± 0.3	35 790 ± 40	586.0 ± 3.0	17 060 ± 85
287.5 ± 0.4	34 780 ± 45	685.0 ± 4.5	14 600 ± 90
333.7 ± 0.6	29 970 ± 55		
360.9 ± 0.8	27 710 ± 60		
418.4 ± 1.1	23 900 ± 70		
453.2 ± 1.4	22 070 ± 75		
536.2 ± 2.3	18 650 ± 85		
637.5 ± 3.8	15 690 ± 90		

13.3.3 *Absorbance calibration*

An absorbance standard is a substance having a known optical density at a given wavelength. It can be either a solid or a solution, and a wide range of both types are reviewed in Volume 1 ([3], pp. 48–93, 135–6). The most accurate, reliable and convenient standards are metal-on-glass neutral density filters, though they must be handled with extreme care. These can be purchased as a set of several filters covering a range of optical densities, and are either calibrated by one of the standards authorities, such as the National Physical Laboratory (NPL) or National Bureau of Standards (NBS), or are calibrated by the manufacturer against a NPL or NBS standard. These filters may change slightly with age and with handling, and should be periodically returned for re-calibration, say, once every year.

The supplier or standard authority will detail conditions for the use of the filters. Since the filters act by reflecting the light that they do not transmit, anomalous readings can be obtained when they are used with some instruments because of multiple reflections of the measuring beam between the filter and other optical components. These inter-reflection errors can be diagnosed by placing the filter at a slight angle to the measuring beam: if there is no significant change in the measured absorbance, then no inter-reflection is present. Verrill [4] discusses this problem in detail.

While neutral density filters provide a convenient check on the instrument, the use of solution standards checks the user's capabilities as well. The most commonly used standard is potassium dichromate in 0.005 M sulphuric acid. This can be purchased ready-made, or can be prepared as follows. Analytical grade potassium dichromate is dried for 1 h at 110°C. Two solutions in 0.005 M analytical grade sulphuric acid are prepared. Solution A containing 50 ± 0.5 mg in 1 litre is used for the absorbance range 0.2–0.7 A; solution B containing 100 ± 1 mg in 1 litre is used for the range 0.4–1.4 A.

Measurements should be made with 10 mm pathlength cells with the temperature controlled within the range 15–25°C using 0.005 M sulphuric acid as the reference. Table 13.2 gives the accepted values for the solutions at two maxima and two minima. The tolerances represent acceptable limits based on the uncertainties of the literature values, the temperature coefficient of the molar absorptivities of potassium dichromate over this temperature range, and the degree of human error normally expected in an analytical laboratory. Fig. 13.3

Table 13.2: *Absorbance values and tolerances for selected maxima and minima for acid potassium dicromate solutions. Taken from Burgess and Knowles [3]. See also Fig. 13.3.*

Wavelength	Absorbance	
(nm)	Solution A	Solution B
235 (min.)	0.626 ± 0.009	1.251 ± 0.019
257 (max.)	0.727 ± 0.007	1.454 ± 0.015
313 (min.)	0.244 ± 0.004	0.488 ± 0.007
350 (max.)	0.536 ± 0.005	1.071 ± 0.011

gives reference spectra for potassium dichromate measured under both acid and alkaline conditions.

This routine only checks the absorbance scale at four points, and a more comprehensive check on the instrument's linearity is more difficult. If a solution standard is selected that obeys Beer's Law, serial dilutions can be measured in a single cell. Potassium dichromate is not recommended for this, but other standards are given in

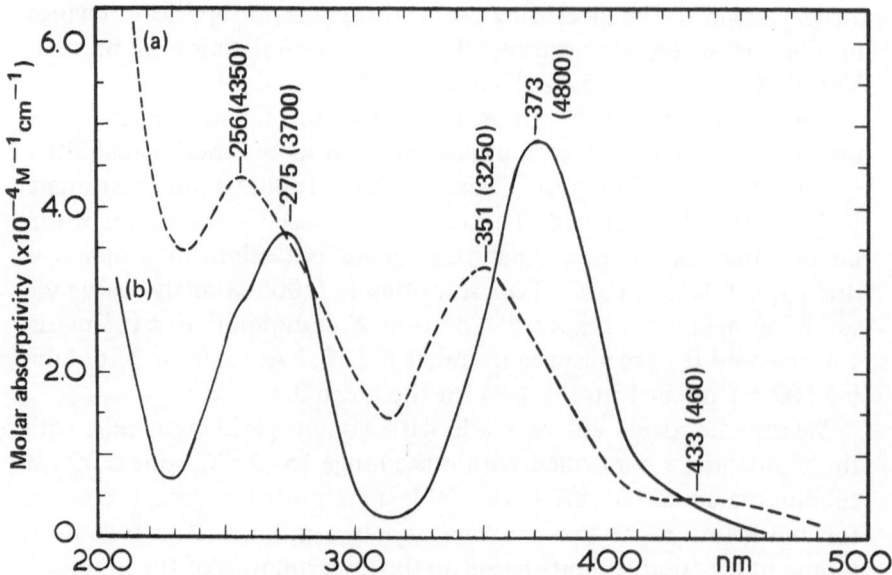

Fig. 13.3 *Absorbance spectra of (a) potassium dichromate in acidic solution (---); and (b) potassium chromate in 5 mM potassium hydroxide (——). The latter is effectively identical with the spectrum of potassium dichromate measured under the same conditions. Redrawn from Perkampus et al. [6].*

Volume 1 ([3], pp. 48–93, 135–6). Appreciable errors can be introduced by the volumetric manipulations, and the alternative approach may be more satisfactory, in which a single solution is measured in a series of cells ranging in pathlength from 5 to 40 mm. The precision of the pathlength of the cells will undoubtedly be greater than that of the dilutions. However, the results obtained from such tests must be interpreted with caution as Clarke [5] points out, and are liable to underestimate deviations from linearity.

13.3.4 *Stray-light checks*

Measuring the total stray-light characteristics of a spectrometer is extremely time-consuming, and it is normally sufficient to measure the instrumental stray-light (ISL), that is, the overall effect of stray-light in the instrument manifested as deviations in the absorbance read-out. However, it must be borne in mind that the effect of stray-light on measured absorbance will be very much dependent on the type of sample being measured. Consequently, this check will serve to monitor the deterioration of the instrument, but it will not be possible to relate your measurement to the manufacturer's specification. Older spectrometers are liable to have a high stray-light characteristic which varies with time and thus require regular checks. A modern instrument should have very low stray-light, and this should remain so. A six-monthly check would not be an unreasonable burden on the user.

The check is performed by using a liquid or solid filter with a sharp 'cut-off', that is, a filter which transmits well at higher wavelengths, but whose transmittance falls rapidly at shorter wavelengths and is effectively opaque from this point down to the shortwave limit of the UV region. Table 13.3 gives a list of liquid filters, and the spectral

Table 13.3: *Cut-off filters for stray-light tests, taken from the ASTM standard [7].*

Spectral range (nm)	Filter liquid	Pathlength (mm)
165–173.5	Water	0.10
170–183.5	Water	10.0
175–200	Aqueous KCl (12 g l^{-1})	10.0
195–223	Aqueous NaBr (10 g l^{-1})	10.0
210–259	Aqueous NaI (10 g l^{-1})	10.0
250–320	Acetone	10.0
300–385	Aqueous $NaNO_2$ (50 g l^{-1})	10.0

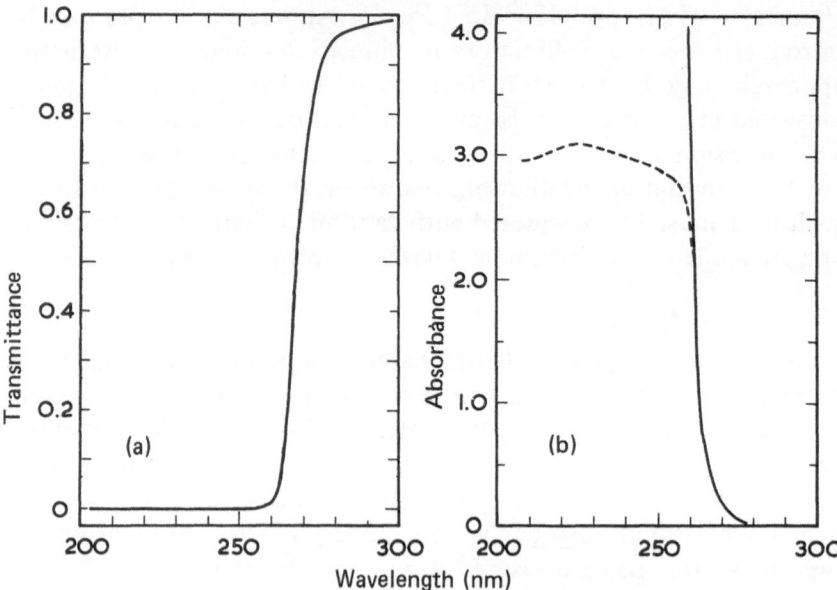

Fig. 13.4 *Spectra for an aqueous solution of sodium iodide (10 g l⁻¹) meas-ured in a 10 mm cell. (a) Transmission spectrum; (b) true absorption spectrum (——) and apparent absorbance (– – –) measured on a typical laboratory spectrometer. The apparent absorbance of 3.08 A at 225 nm means that when the monochromator is set to that wavelength, 0.08% of the radiation leaving the exit slit is of wavelengths greater than about 265 nm.*

range of their cut-off. Fig. 13.4a gives the transmission spectrum of one of these solutions and shows that the transmittance of the solution is less than 0.001 over its working range. Fig. 13.4b shows the true absorbance spectrum of the solution and a typical spectrum measured with a popular spectrometer with a grating monochromator. Each of the test solutions can be used to check the stray-light that is present when the monochromator is set to wavelengths below the cut-off range. Put the solution in the correct pathlength cell and scan the instrument down from a wavelength above the cut-off. When the cut-off is reached, the absorbance should increase rapidly to the maximum value that the instrument can indicate, and remain at that value at all shorter wavelengths. However, if stray-light is present, the instrument will not reach the maximum value, or may reach a high value and then start to fall again, giving an asymmetrical peak. If, for example, the apparent absorbance at a wavelength λ that is below the cut-off region is 2 A, then 1% of the measuring beam that is falling on the cell is reaching the detector: we know the sample is opaque at this wavelength and so either the light is passing through the walls of the cell or over the surface of the solution, or 1% of the radiation leaving the monochromator is of wavelengths greater than the cut-off, and is therefore not absorbed by the solution. The first possibility can be checked by visual inspection. The consequence of 1% stray-light is that when the instrument is used to measure absorbance at λ, the reading will be too low; for example, a sample of $A = 1.0$ will have an apparent absorbance of 0.959, i.e. 4% low.

A Corning Vycor glass filter provides a rapid method of checking stray-light for monochromator settings in the range 200–210 nm. It is particularly useful because this is a wavelength region where stray-light becomes a serious problem. Unfortunately, this filter is not suitable as a standard because the cut-off edge is not sufficiently steep (see Appendix A.2.1).

Stray-light is an insidious problem and it is not always easy to decide when it has become too high. The manufacturer's literature will give limits for an instrument at specified wavelengths, and if you find that your measurements exceed these limits, consider cleaning the optical components of the monochromator. You may decide to tackle the job yourself (see Section 13.5), or it may be prudent to call in the service engineer. Consultation of earlier check sheets will show whether the problem is one of progressive contamination or the catastrophic failure of some component. A full discussion of stray-light will be found in Volume 1 ([3], pp. 94–110, 135–6).

13.3.5 *Lamp output*

If your instrument has a slitwidth or energy indicator, this can be used to check the output from the source lamps. Routine measurement at one or more wavelengths in the wavelength range of each lamp, for example 220 nm and 500 nm, and plotting the results on a control chart will show up lamp deterioration and help to predict lamp failure.

13.3.6 *Realistic performance criteria*

The manufacturer's specification will give figures for stray-light, photometric accuracy and wavelength accuracy which will serve as guidelines for the performance that you can expect from the instrument. However, these figures were obtained with a new instrument operating under the best environmental conditions, and it is unlikely that your five year-old instrument can reach this performance. While one can attempt to maintain the instrument in 'showroom' condition, it is probably more realistic to set personal tolerances based on the level of performance that you require from the instrument. As soon as the checks show that the performance has fallen outside these tolerances, action must be taken.

Data from the check sequence outlined above should give an insight into results obtained with the instrument under 'normal' operating conditions. For non-routine measurements, more stringent tests must be worked out that are relevant to the experiments to be performed.

10.4   Simple fault finding

A regular check routine not only validates your results, but aids fault diagnosis and the distinction between instantaneous and insidious faults. If the instrument shows an apparent malfunction during use, do not immediately assume that it is at fault. First check that all the controls are set correctly. Even the most experienced operator can forget to switch on the deuterium lamp. Next check that the experimental conditions do not prevent the spectrometer from functioning properly. Remove the cells and see whether the 'fault' is still present.

If the problem seems to lie within the spectrometer, see whether the fault is related to a particular wavelength setting and is thus probably caused by a lamp or filter change. On the other hand, if it is related to the particular absorbance reading, it may be due to a bad contact on a potentiometer or recorder slide wire. Turn the wavelength setting to 550 nm, open the slits and look at the light going through the cell. Remove the cover of the lamp house and inspect the condition of the lamps and mirrors. Decide whether the fault is likely to be optical, electronic or mechanical. Again, visual inspection may reveal a mechanical fault. Lubrication must be carried out with extreme caution: use only very small amounts of oil when essential, and never use an aerosol lubricant or cleaning spray, for oil vapour can seriously affect optical components, possibly irreversibly. Electrical fault finding requires a greater degree of expertise, but first check that all plugs, connectors and printed circuit boards are making proper contact. Check all potentiometers and switches particularly if the instrument has been out of use for some time. Check the fuses. The advent of self-diagnostic circuit boards will aid electronic fault finding in the future.

Careful investigation should be made before calling out the service engineer, for he will not appreciate being called to an instrument that could have been rendered operational by the user, and his time is very costly. A service contract should be considered, although these can be very expensive and the service offered by different companies varies greatly in its quality and coverage, and hence its usefulness to you. Never attempt any repairs yourself unless you are fully aquainted with the service manual and the safety precautions. Ensure that a comprehensive supply of spares such as fuses and lamps is readily available.

13.5   Cleaning optical components and lamps

Modern optical components are often coated with a protective layer that reduces the rate of contamination and, when cleaning becomes necessary, allows more rigorous cleaning procedures than can be used with the optics of older instruments. All optical components must be handled with extreme care, and cleaning should only be attempted by someone who is familiar with the procedures involved. It is unwise to attempt to clean diffraction gratings, or even to touch them. Lenses, mirrors, prisms and lamps can be cleaned by the following procedure:

1. Never touch the optical surfaces with any solid substance.
2. Wash with a stream of warm dilute detergent, e.g. household washing-up liquid.
3. Rinse with a stream of distilled water.
4. Rinse with a stream of absolute ethanol or methanol, and leave to dry.
5. If still contaminated, repeat the washing sequence.
6. If it is impossible to remove stubborn contamination, repeat the detergent wash and gently mop the surface with lens tissues or lint-free cloth.
7. Take care that grease is not transferred from the fingers on to the optical surfaces during the washing routine, and that no fingerprints are left on the components when they are reassembled.
8. Wipe source lamps with lens tissues soaked in ethanol and held in forceps before switching on, since fingerprints can be permanently etched on to the hot lamp envelope.

References

1 Hughes, D. (1978), *Hazards of Occupational Exposure to Ultraviolet Radiation*, University of Leeds Industrial Services, Leeds.

2 Edisbury, J.R. (1966), *Practical Hints on Absorption Spectrometry*, Hilger and Watts, London.
3 Burgess, C. and Knowles, A. (Eds) (1981), *Techniques in Visible and Ultraviolet Spectrometry*, Vol. 1, *Standards in Absorption Spectrometry*, Chapman and Hall, London.
4 Verrill, J.F. (1983), *UV Spectrom. Grp Bull.*, **11**, part 1. In press.
5 Clarke, F.J.J. (1981), *UV Spectrum. Grp Bull.*, **9**, 81.
6 Perkampus, H., Timmons, C.J. and Sandeman, I. (Eds) (1971), *UV Atlas of Organic Compounds*, Verlag Chemie, Weinheim; Butterworths, London.
7 ASTM Standard E-387 (1978), *Annual Book of ASTM Standards, 1978*.

Appendices

A1 Solvent characteristics

A1.1   Properties of common solvents

Data from [1, 2] and various other sources.

Solvent	MW	MP (°C)	BP (°C)	Density (kg l⁻¹) at 20°C	Density gradient[e] (kg l⁻¹ °C⁻¹)	RI[f]	Hazard[g]	Solubility in water[h]	Polarity[i]	Cut-off wavelength[j] (nm)	Spectrum
Acetone	58	−94.8	56.2	0.791	–	1.3591	F	∞	20.7	331	A1.2
Benzene	78	5.5	80.1	0.8790	0.001 05	1.5011	F, T	0.07	2.3	280	A1.4
Chloroform[a]	119	−63.5	61.2	1.489	0.001 86	1.4460	H	0.82	4.8	246	A1.3
Cyclohexane	84	6.6	80.9	0.7787	0.000 92	1.4263	F	0	2.0	211	A1.4
Dimethyl-formamide	73	−61	153	0.9445[d]	0.000 95	1.4269	H	∞	36.7	270	A1.2
Dimethyl sulphoxide	78	18.5	189	1.1014	0.000 8	1.4783	H	∞	–	270	A1.3
Dioxan	88	11.8	101.3	1.0337	0.001 12	1.4224	F, H	∞	2.2	215	A1.4
Ethanol[b]	46	−117.3	78.3	0.7893	0.000 85	1.3610	F	∞	24.3	207	A1.3
Hexafluoro-propan-2-ol	168	−3.4	38.2	1.59[d]	0.003	1.2752	H	∞	–	<195	A1.4
Hexane[c]	86	−95.3	68.7	0.649	–	1.3749	F	0	1.9	199	A1.2
Methanol	32	−97.8	64.6	0.7913	0.000 934	1.3288	F	∞	32.6	210	A1.2
Water	18	0	100	0.9982	0.000 21	1.3330	–	∞	78.5	<195	–

a Commercial samples generally contain ethanol.
b 100% (absolute).
c Generally sold as 'Hexane fraction', boiling range 67–70°C
d At 25°C.
e Change in density per °C at ambient temperature.
f Determined at 20°C using 589.3 nm light.
g F: highly flammable, flash point < 32°C; T: toxic substances; H: harmful substances.
h Parts per hundred at ambient temperature.
i Dielectric constant at ambient temperature.
j Wavelength at which transmittance of a 10 mm layer has fallen to 0.25, representing lower limit of useful range.
Abbreviations: MW, molecular weight; MP, melting point; BP, boiling point at atmospheric pressure; RI, refractive index.

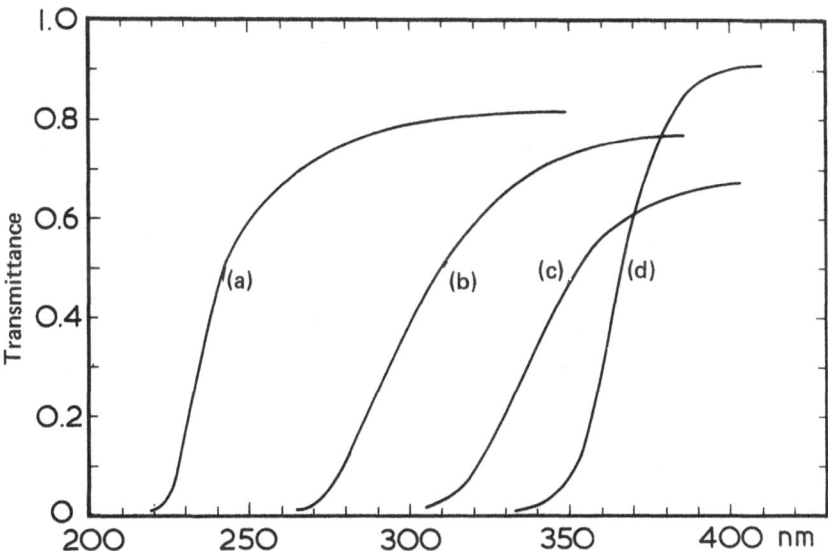

Fig. A1.2 *Transmission spectra for solvents measured in a 10 mm cell against a water reference. (a) Hexane, (b) methanol, (c) dimethylformamide, (d) acetone. Redrawn from Perkampus et al. [3].*

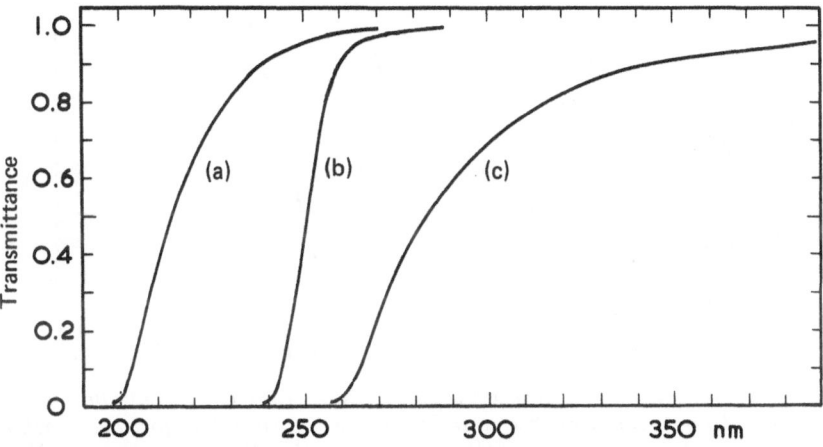

Fig. A1.3 *Transmission spectra for solvents measured in a 10 mm cell against a water reference. (a) Ethanol, (b) chloroform, (c) dimethylsulphoxide. Redrawn from Perkampus et al. [3].*

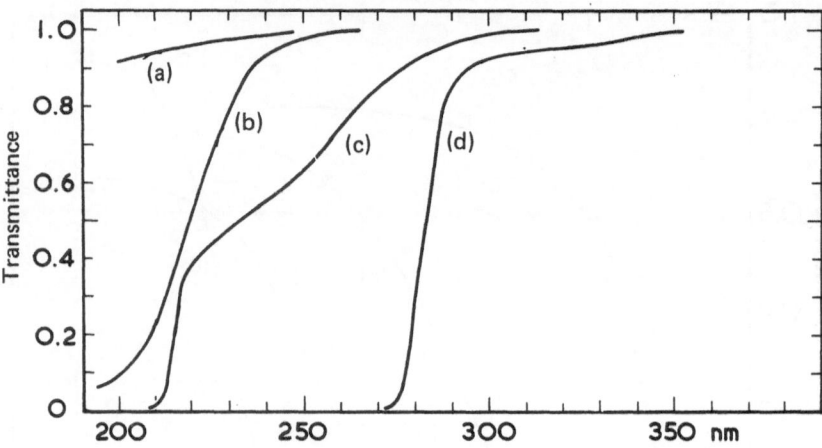

Fig. A1.4 *Transmission spectra for solvents measured in a 10 mm cell against a water reference. (a) Hexafluoropropan-2-ol. Redrawn from BDH Handbook [2], (b) cyclohexane, (c) dioxan, (d) benzene. Redrawn from Perkampus et al. [3].*

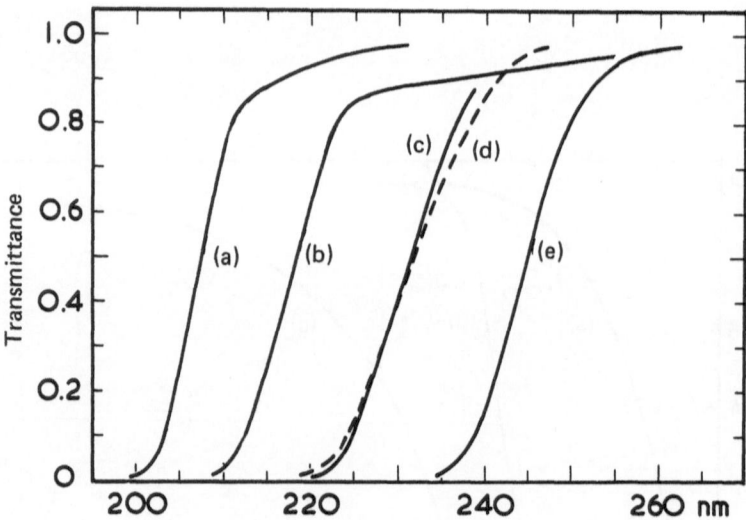

Fig. A1.5 *Transmission spectra of buffer solutions at 0.1 M concentration measured in a 10 mm cell against a water reference. (a) Disodium hydrogen orthophosphate—sodium dihydrogen orthophosphate, pH 6.85, (b) tris buffer, pH 7, (c) sodium carbonate—sodium hydrogen carbonate, pH 9.9, (d) sodium acetate—acetic acid, pH 4.6, (e) sodium citrate—citric acid, pH 4.6. Redrawn from Perkampus et al [3].*

A2   Transmission of optical materials

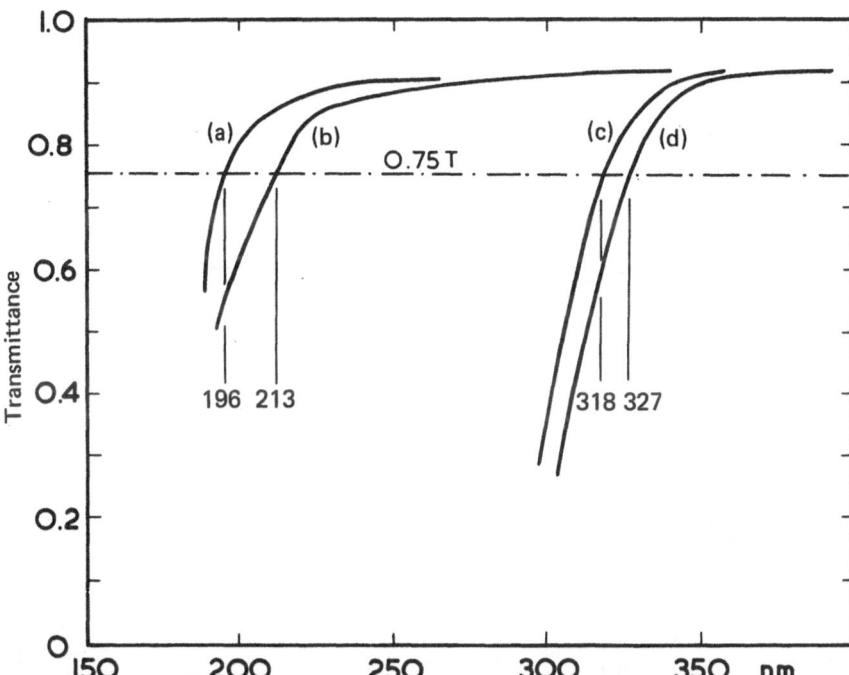

Fig. A2.1 *Transmission spectra for 10 mm normal cells filled with freshly distilled water measured against air. (a) Fused synthetic silica, (b) fused quartz, (c) 'special optical' glass, (d) glass. The line at 0.75 T indicates the recommended shortest wavelength at which the cells should be used. Redrawn from Burgess and Knowles [4].*

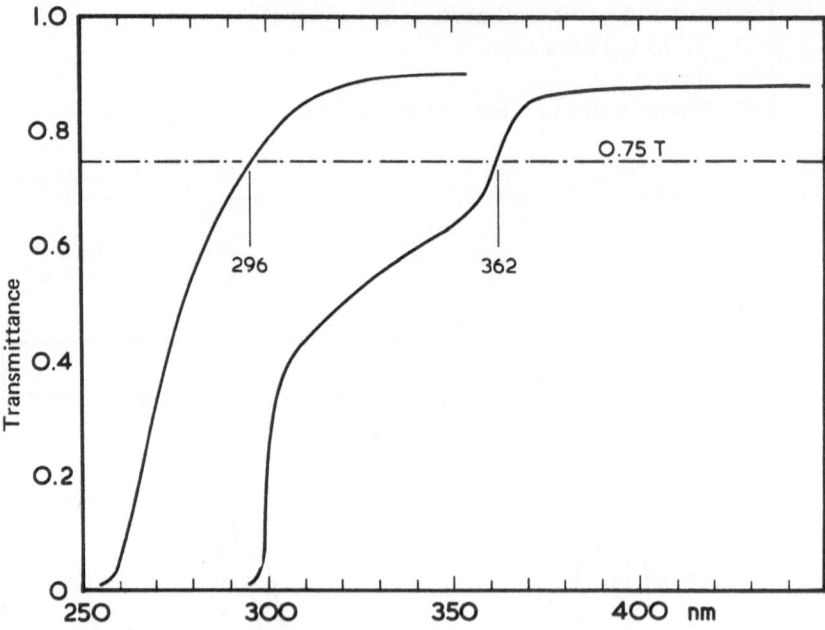

Fig. A2.2 *Transmission spectra for two 10 mm plastic cells supplied by different manufacturers. The line at 0.75 T indicates the recommended shortest wavelengths at which the cells should be used.*

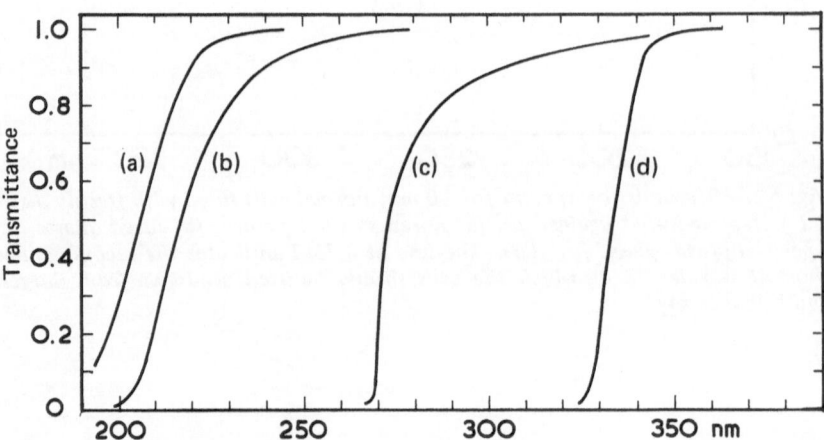

Fig. A2.3 *Transmission spectra of various optical materials in the specified thickness. (a) Corning 9-54 glass filter (Vycor), 2.0 mm; (b) Corning 0-53 (Pyrex), 4.0 mm; (c) Chance–Pilkington HA3 glass filter (heat absorbing), 4.0 mm; (d) ICI Perspex acrylic plastic, 6.2 mm.*

A3 Wavelength standards

A3.1 Holmium ions in solution
A3.2 Samarium ions in solution
A3.3 Holmium glass
A3.4 Didymium glass

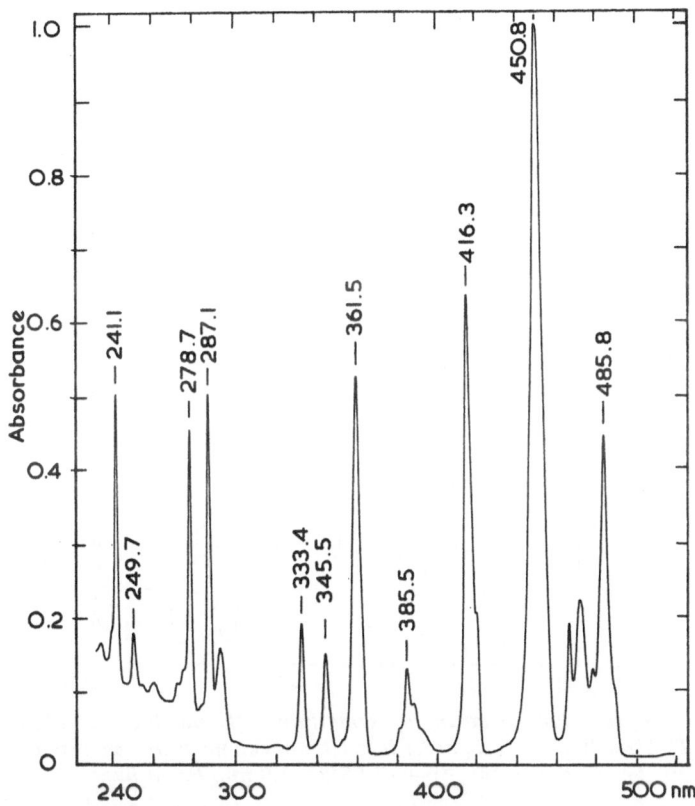

Fig. A3.1 *Absorption spectrum of holmium (III) perchlorate solution (10% w/v in 17.5% w/v perchloric acid) in a 10 mm cell measured against a perchloric acid blank. Scanned at 1 nm s^{-1} with SSW = 1.0 nm. The peaks at 241.1, 287.1 and 361.5 nm are recommended by the European Pharmacopoeia [5] for wavelength calibration. From Burgess [6].*

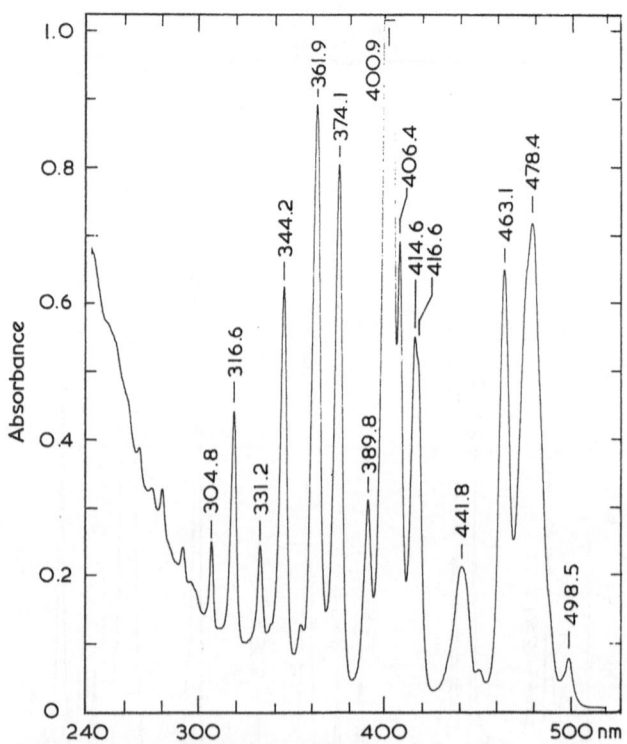

Fig. A3.2 *Absorption spectrum of samarium (III) perchlorate solution (10% w/v in 17.5% w/v perchloric acid) in a 10 mm cell measured against a perchloric acid blank. Scanned at 1 nm s^{-1} with SSW = 1.0 nm. From Burgess [6].*

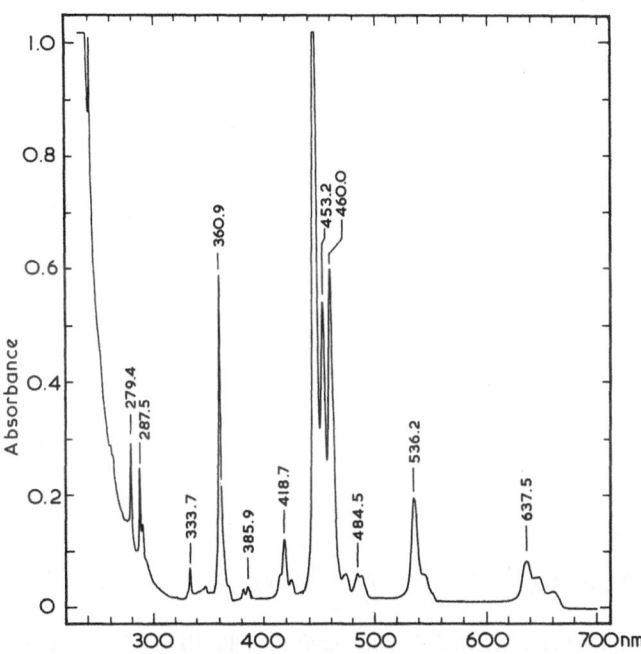

Fig. A3.3 *Absorption spectrum of a holmium glass filter of unknown origin, scanned at 1 nm s⁻¹ with SSW = 0.2. The peak values are taken from McNierney and Slavin [7]. Note that there are variations in peak positions between different batches of glass.*

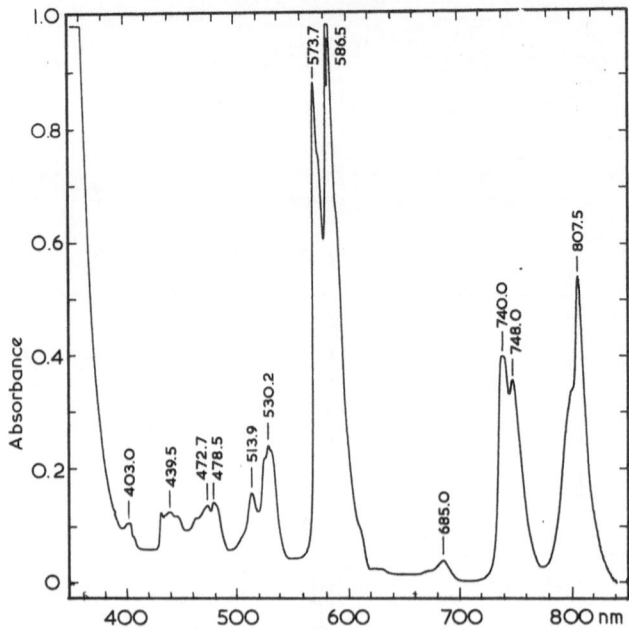

Fig. A3.4 *Absorption spectrum of a didymium glass filter (Chance–Pilkington ON 12, 2.0 mm thick), scanned at 1 nm s^{-1} with SSW = 0.2. The peak values are taken from Perkampus et al. [3]. Note that there are variations in peak positions between different batches of glass.*

References

1 *Materials for UV and Visible Spectroscopy* (1983), BDH Chemicals Ltd, Poole.

2 *Laboratory Chemicals and Biochemicals* (1981), BDH Chemicals Ltd, Poole.

3 Perkampus, H., Sandeman, I. and Timmons, C.J. (1971), *UV Atlas of Organic Compounds*, Butterworths, London; Verlag Chemie, Weinheim.

4 Burgess, C. and Knowles, A. (Eds) (1981), *Techniques in Visible and Ultra-violet Spectrometry*, Vol. 1, *Standards in Absorption Spectrometry*, Chapman and Hall, London.

5 *European Pharmacopoeia* (1980), 2nd Edn, Section V.6.19, Maisonneuve, Sainte Ruffine, France.

6 Burgess, C. (1977), *UV Spectrom. Grp Bull.*, **5**, 77.

7 McNierney, J. and Slavin, W. (1962), *Appl. Optics*, **1**, 365.

Index

The definition of many terms and synonyms will be found in the Glossary, which appears on pp. xiii to xxii.